Springer Series in

MATERIALS SCIENCE 124

Springer Series in

MATERIALS SCIENCE

The Springer Series in Materials Science covers the complete spectrum of materials physics, including fundamental principles, physical properties, materials theory and design. Recognizing the increasing importance of materials science in future device technologies, the book titles in this series reflect the state-of-the-art in understanding and controlling the structure and properties of all important classes of materials.

Please view available titles in *Springer Series in Materials Science* on series homepage http://www.springer.com/series/856

David H. Kirkwood
Michel Suéry
Plato Kapranos
Helen V. Atkinson
Kenneth P. Young

Semi-solid Processing of Alloys

With 103 Figures

Springer

Dr. David H. Kirkwood
Dr. Plato Kapranos
University of Sheffield
Department of Engineering Materials
Mappin Street, Sheffield S1 3JD, UK
E-mail: d.h.kirkwood@sheffield.ac.uk
p.kapranos@sheffield.ac.uk

Professor Helen V. Atkinson
University of Leicester,
Department of Engineering
University Road, LE2 3TJ Leicester, UK
E-mail: hva2@leicester.ac.uk

Dr. Michel Suéry
Grenoble INP
Laboratorie SIMAP
Groupe GPM2
BP 46, 38402 St.-Martin d'Hères, France
E-mail: michel.suery@simap.grenoble-inp.fr

Dr. Kenneth P. Young
Vforge Inc.
5567 West 6th Ave, Lakewood, CO 80214, USA
E-mail: kyoung@vforge.com

Springer Series in Materials Science ISSN 0933-033X
ISBN 978-3-642-00705-7 e-ISBN 978-3-642-00706-4
DOI 10.1007/978-3-642-00706-4
Springer Heidelberg Dordrecht London New York

Library of Congress Control Number: 2009929024

Cover design: SPi Publisher Services

Printed on acid-free paper

Springer is part of Springer Science+Business Media (www.springer.com)

Foreword

Semisolid metallurgy (SSM) is now some 37-years-old in terms of time from its conception and first reduction to practice in the laboratory. In the intervening years, there has been a steadily growing body of research on the subject and the beginning of significant industrial applications.

The overall field of SSM comprises today a large number of specific process routes, almost all of which fall in the category of either "Rheocasting" or Thixocasting." The former begins with liquid metal and involves agitation during partial solidification followed by forming. The latter begins with solid metal of suitable structure and involves heating to the desired fraction solid and forming.

Research over the past 37 years, and particularly over the last decade, has provided a detailed picture of process fundamentals and led to a wide range of specific SSM processes and process innovations. Industrial studies and actual production experience are providing a growing picture of the process advantages and limitations.

At this time, the conditions for eventual wide adoption of SSM appear favorable, both for nonferrous and ferrous alloys. It must, however, be recognized that major innovations, such as SSM become adopted only slowly by industries where capital costs are high, profit margins are modest, and failure to meet customer commitments carries a high penalty.

As demonstrated by the references found in chapters of this book, enthusiasm of SSM researchers throughout the world is high, and industrial successes are beginning to be evident. The newness of the technology and the continuing development of our process understanding make it evident that key to increasing rate of industrial utilization of SSM will be (1) close collaboration between researchers in the field and production personnel, and (2) patience and persistence in the industrial development phase.

Cambridge, MA, November 2009 *Merton C. Flemings*

Preface

Semisolid processing originates with the research at MIT on the viscosity behaviour of tin/lead alloys during solidification, published by Spencer, Mehrabian, and Flemings in 1972.The original purpose was to examine the shear stress of semisolid structures with a view to shedding light on hot tearing mechanisms in solidifying alloys, but it was discovered that the process of shearing clearly altered the microstructure, changing the dendritic morphology of the solid into more spherical form that brought about a dramatic fall in shear stress and viscosity. This is of course the phenomenon of "Thixotropy," well known to ceramicists and polymer scientists, but until then not observed in metallic systems. This discovery at MIT initiated an in-depth study of thixotropy in alloys, and the awareness that stirred semisolid alloys could be injected as in die casting into molds with certain advantages: The controlled higher viscosity flow meant that less air was entrapped in the casting and that the finer equiaxed grains formed on solidification, both would result in better mechanical properties. The work carried out in the last 40 years in industry and laboratories around the world and published in the Biennial International Conference Proceedings on Semisolid Processing of Alloys and Composites, as well as in academic journals, has borne out these expectations: There is now a thriving industry in the production of thixoformed components in aluminum and magnesium alloys for automotive components, and also in the electronics and telecommunications industries. However, this is still a small proportion of all castings produced by more conventional processes, and this restriction is mainly due to the cost of the preformed thixoformed billet and the inability to recycle scrap in-house. Both these problems could be overcome by the recent development of the new slurry generating technologies permitting rheocasting (that is integrating the formation and injection of slurry on site, see Industrial Chap. 10), obviating the need to partially remelt a preformed slug and providing important production savings – thus allowing direct competition with conventional high pressure die casting.

The subject matter in this monograph is discussed in three parts. The first part (A) deals with the microstructure of semisolid slurries resulting from nucleation and dendrite fragmentation (Chap. 1), and the changes that occur during subsequent isothermal heat treatment (Chap. 3). This is followed in Chap. 4 by an account of the more recent advances in generating slurries in-house, thus allowing significant economies in component production. The rheology of alloy slurries is examined in Chap. 6 and the modeling of semisolid flow is outlined in Chap. 7, providing a tool to optimise die design and prevent defect formation. Finally the industrial applications of semisolid processing are dealt with in Chap. 10, in which different techniques are discussed for reheating and partially melting preformed billets, followed by isothermally heat treating them before thixoforming. Chapters 11and 12 describe the practical considerations of process control and die design, and the empirical rules that have been established as a guide to produce sound castings on a regular basis.

I wish to acknowledge the efforts of the authors in providing a balanced, self-contained, and up-to-date account of semisolid alloy processing that follows. It must be admitted that this monograph was begun some years ago but has been delayed in completion by the remarkable progress in the industrial technology. Hopefully with the advent of new rheocasting technologies, one phase in development has been completed – that is for light alloys. Copper based alloys for the production of intricate shapes must now be an achievable goal and this should stimulate the development of semisolid forming of high melting point alloys, such as stainless steels, nimonics, and titanium alloys using special die materials.

I should especially express my thanks to Dr Plato Kapranos, my colleague of many years, for his general help – not least in the area of the computer manipulation of text, and to Professor Merton C. Flemings, the father of semisolid processing, for providing the Preface and his support for the project, and also to Professor Hans Warlimont for suggesting it in the first place and for his continuous encouragement.

Sheffield *David Kirkwood, Plato Kapranos*
Grenoble *Michel Suéry*
Leicester *Helen V. Atkinson*
Lakewood *Kenneth P. Young*
November 2009

Contents

Part II Rheology and Modeling
M. Suery and H. Atkinson

Part III Industrial Applications of Semisolid Processing
Ken Young and Plato Kapranos

Part I
Evolution and Design of Microstructure in Semisolid Alloys

D.H. Kirkwood

Chapter 1
Fundamental Aspects

1.1 General and Local Equilibrium in Alloy Systems

The original semisolid forming process, developed at MIT in 1972, involved stirring an alloy during solidification to produce a slurry of spheroidal primary particles in a liquid matrix, which was then injected directly into a die to produce a solid component. This was termed "rheocasting." Subsequently, it was found more convenient to solidify the slurry completely during the continuous casting of an electromagnetically stirred strand, which was then cut into slugs for partial remelting back into semisolid billets on demand. These could be loaded into a diecasting machine in this state for injection into the die. This alternative process route is called "thixoforming," and until recently, it was the preferred industrial process. For this reason, the microstructure developed during the reheating and melting for thixoforming will be considered first. Experimentally, it has been found that the most effective fraction solid f_s for thixoforming, lies between 0.5 and 0.6. Below this range, the semisolid slug becomes too soft to support its own weight and sags during remelting; above this range, it is too stiff to flow readily and fill the die. However, the slurry technologies used in rheocasting typically operate at lower fraction solids and rely on the ability to pour the semisolid alloy much like a liquid (see Chap. 4). It has been observed in practice that the fraction solid is in fact a critical factor for effective thixoforming, and therefore, both good temperature control and lack of sensitivity of f_s to small temperature variations of the alloy are essential to efficient manufacturing.

A determination of the change in f_s with temperature in a binary alloy system under strict equilibrium conditions may be obtained by referring to the phase diagram. For an alloy of overall composition C_o, conservation of solute requires that $C_o = C_l f_l + C_s f_s$, where the compositions are expressed in weight percentage and the fraction f as a weight fraction of liquid or solid. It reduces to $f_s = (C_l - C_o) / (C_l - C_s)$ (see Fig. 1.1), where C_s and C_l are the equilibrium compositions of solid and liquid, respectively, at temperature T. This is the familiar lever rule for obtaining the fraction solid; it allows one to calculate f_s where *general equilibrium conditions* prevail, that is, diffusion in the liquid and solid has removed all compositional gradients in the system. In particular that during solidification, the

David H. Kirkwood et al., *Semi-solid Processing of Alloys*,
Springer Series in Materials Science,
DOI 10.1007/978-3-642-00706-4_1,

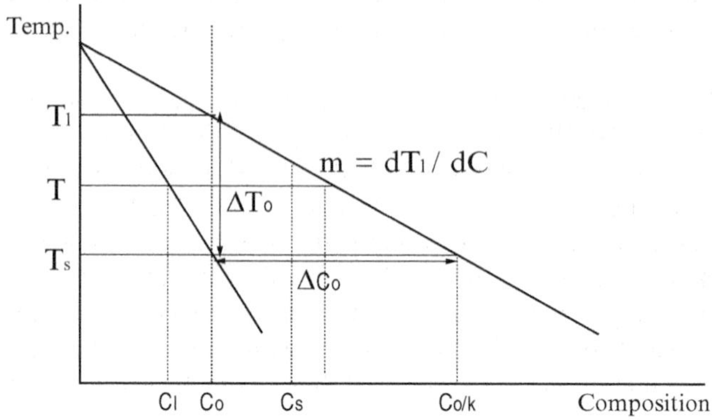

Fig. 1.1 Simple Binary Alloy Equilibrium Diagram

build-up of solute in the solid at the moving interface is continuously removed by back diffusion into the solid. However, a simple calculation of the time needed for diffusion to occur at the center of a 100 μm diameter spherical particle, assuming a typical diffusion coefficient of $D = 10^{-13}\,\mathrm{m^2\,s^{-1}}$, estimates that several hours are required. It is clear that under the fairly rapid cooling conditions employed in (Direct Chill) casting of commercial billets, equilibrium cannot prevail, nor during the soaking period prior to injection into the die.

The concept of *local equilibrium* is very often employed in transformation problems, as in solidification, where by the nature of things *general equilibrium* does not prevail. The assumption here is that the reaction kinetics at the interface (or interfaces) is very rapid (or there is negligible undercooling at the interface) compared with solute or heat transfer to or from the moving interface, and the latter therefore control the kinetics of transformation. Under these conditions, the interface compositions of both the liquid and solid may be obtained from equilibrium considerations, that is, by the use of phase diagrams. With this assumption, together with those of no diffusion in the solid and complete diffusion in the liquid, we obtain the other extreme prediction for f_s given by the Scheil equation: $f_s = 1 - (C_l/C_o)^{-1/(1-k)}$, which predicts a smaller value at a given temperature than the lever rule. Of course, some back diffusion into the solid will occur even for rapid cooling and a plot of f_s for Al–4.5%Cu over different cooling rates estimated by Flemings [1] is given in Fig. 1.2, which shows that Scheil prediction is quite accurate in the region of practical interest for semisolid processing ($f_s \sim 0.5$) and back diffusion has only a minor influence. The Scheil equation may be easily modified to take account the variation of k with temperature, and introducing more solutes, and solved using numerical methods. It should be added that the above predictions refer only to solutes that form substitutional solutions in the solid: in general, interstitial solutes, such as carbon in steels, diffuse very rapidly and for many purposes may be considered to be in *general equilibrium* between the phases.

Equilibrium phase diagrams assume planar interfaces exist between the phases represented, but may be modified to represent high interfacial curvatures. It may

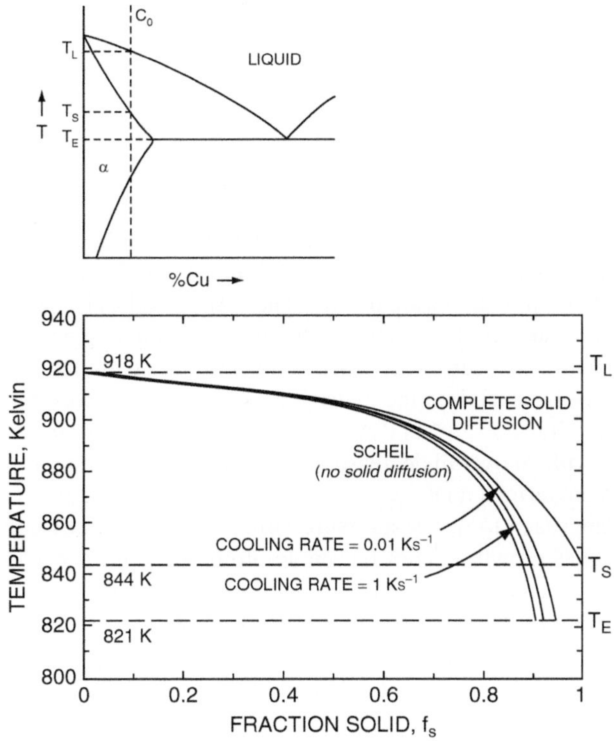

Fig. 1.2 The effect of Cooling Rate on the fraction of Primary Solid formed

be shown [2, 3] that the undercooling produced by interface curvature is given by $\Delta T_r = \frac{K\sigma}{\Delta S_f}$, where σ is the surface energy, ΔS_f the entropy of fusion and $K = 1/r_1 + 1/r_2$, where r_1 and r_2 are the two principal radii of curvature. For metal systems, ΔT_r only becomes important where the radius is less that 10 μm, and will therefore affect processes, such as nucleation, interface instability, dendrite growth, and coarsening.

1.2 Nucleation

1.2.1 Homogeneous Nucleation

The Rate Equation

When a pure metal is cooled below its melting point, the liquid becomes undercooled and is in a metastable state with respect to the solid crystalline phase. There is a thermodynamic barrier to the formation of solid, since the first solid to appear has to create an interface between the two phases that requires the expenditure of energy. We may write for the formation of a small spherical solid embryonic particle of radius r from the liquid melt:

$$\Delta G_r = \frac{4}{3}\pi r^3 \Delta G_v + 4\pi r^2 \sigma,$$

where ΔG_v is the volume free energy change in transforming from liquid to solid, and σ is the interfacial energy between solid and liquid. It is clear that above the melting point both these terms are positive and ΔG_r must be also positive for all values of r and no solid particles can be stable. Below the melting point, ΔG_v becomes negative, and while ΔG_r remains positive for small solid particles (embryos), which will tend to dissolve back into the matrix, larger particles, however, that succeed in growing to the critical size r^* at $\mathrm{d}\,\Delta G_r/\mathrm{d}r = 0$ (see Fig. 1.3) can continue to grow further in size with an overall decrease in free energy of the system. Those particles larger than r^* are termed nuclei.

A simple approach to derive an expression for *homogeneous* nucleation rate I is to assume that growth occurs by single atom addition to nuclei and that the nucleation flux is given by the rate at which the critical embryos r^* grow by one atom jumping across the interface: $I = \nu S^* C^*$, where ν is the jump frequency of an atom, S^* is the number of atoms at the interface, and C^* is the number of critical embryos. This "equilibrium" number of critical embryos, C^*, may be estimated from the ideal entropy of mixing of embryos and single atoms, and the rate of atom transfer from diffusion of atoms in the liquid, to provide an estimate of nucleation rate:

$$I \approx 10^{39} \exp\left(-\frac{\Delta G^*}{kT}\right) \quad \mathrm{s}^{-1}\mathrm{m}^{-3}. \tag{1.1}$$

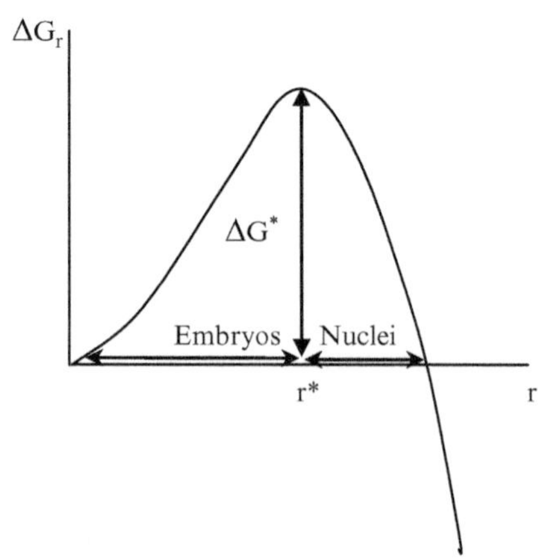

Fig. 1.3 Free energy of growing embryo/nucleus

The value of ΔG^*, the free energy barrier, in (1.1) depends on the undercooling ΔT below the melting point, and is directly proportional to $1/\Delta T^2$. Since this appears within the exponential term, the nucleation rate becomes extremely sensitive to the undercooling ΔT.

Using appropriate values to calculate ΔG^* for aluminum [2,4] in (1.1), the nucleation rate has been plotted against undercooling in Fig. 1.4. For a nucleation rate of $10^6\,\text{s}^{-1}\,\text{m}^{-3}$, which is equivalent to one nucleus appearing per second in a volume of $1\,\text{cm}^3$ being the point at which nucleation can sensibly be observed under experimental conditions, this will occur from the figure at $\Delta T = 190\,\text{K}$, close to the experimental value given below. Figure 1.4 also demonstrates that over a further temperature decrease of 3 K, the rate increases by tenfold, illustrating the extreme sensitivity of nucleation to undercooling. Another way of expressing this sensitivity is to note that at 187.4 K undercooling, it would take 10 s for one nucleus to

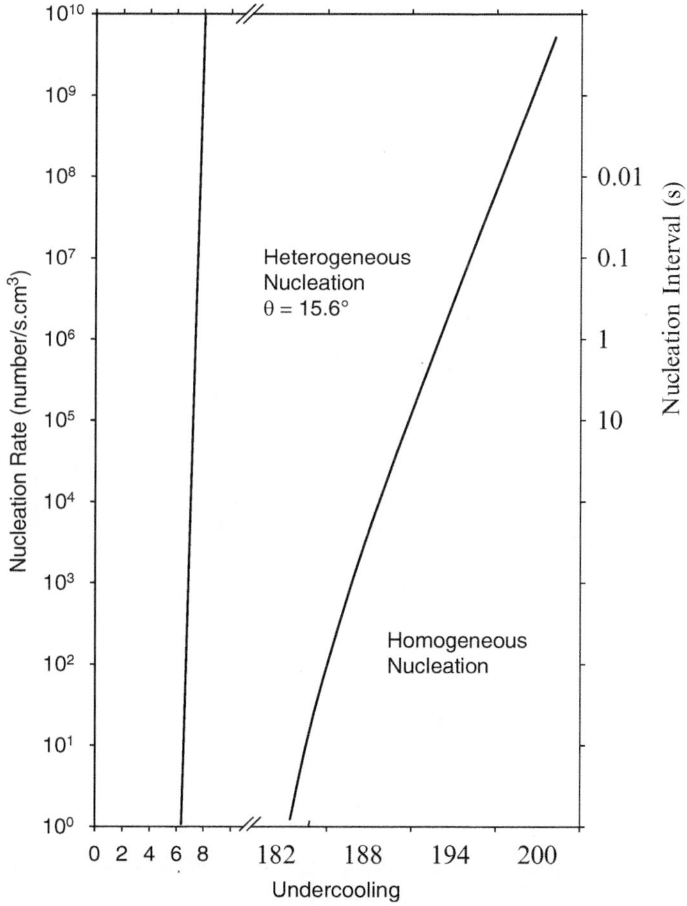

Fig. 1.4 Homogeneous and heterogeneous nucleation rates as a function of undercooling

appear in a melt of 1 cm^3, whereas at 193 K it takes 0.1 s; effectively zero to copious nucleation over a temperature range of 6 K.

Holloman and Turnbull [4] carried out the classical work on homogeneous nucleation in pure metals, avoiding the problem of heterogeneous nucleation on impurity surfaces by dispersing the metal into very fine droplets whose numbers far exceeded the impurity particles present, and found that extremely high undercooling is required to achieve homogeneous nucleation: 195 K for aluminum and 295 K for iron, for example. From these results, they were able to derive the surface energies, which in some cases have been shown to be in good agreement with values derived using different techniques.

1.2.2 Heterogeneous Nucleation

It is clear then that homogeneous nucleation will occur only at very high undercooling in a melt that can never be achieved in normal industrial or even most laboratory solidification conditions. In fact solidification is commonly observed to start in metals within a few degrees of undercooling, and this strongly suggests that nucleation occurs on the surface of impurity particles (*nucleants*) present within the melt, or at the surface of the container, which helps to lower the thermodynamic energy barrier to nucleus formation, giving rise to *heterogeneous* nucleation. This has been treated theoretically [2, 3] by considering the formation of a small cap of solid upon the planar surface of the impurity having a contact angle θ (Fig. 1.5).

Balancing the surface tension forces along the surface, we have:

$$\sigma_{\text{nl}} = \sigma_{\text{sn}} + \sigma_{\text{sl}} \cos\theta.$$

Following a similar procedure used to calculate homogeneous nucleation, the heterogeneous nucleation rate can be determined approximately to be:

$$I_{\text{het}} \approx 10^{30} \exp\left(-\frac{\Delta G^*}{kT} f(\theta)\right) \text{ s}^{-1}\,\text{m}^{-3}, \tag{1.2}$$

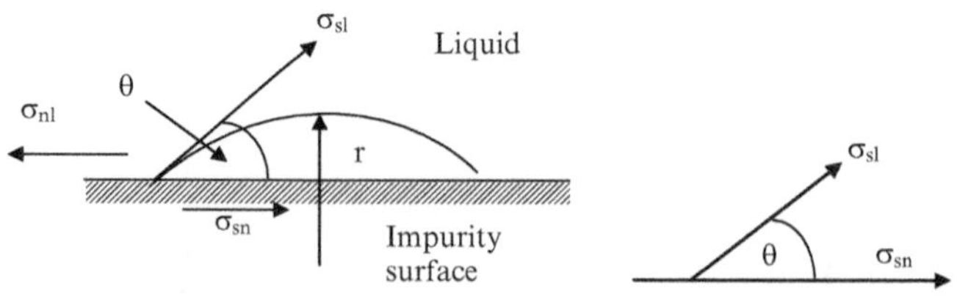

Fig. 1.5 Cap of solid phase forming from the liquid on an impurity surface by heterogeneous nucleation

where the reduced pre-exponential term takes some account of the potential sites on the nucleant surface available for nucleation, and the term $f(\theta) = (2 + \cos\theta)(1 - \cos\theta)^2/4$ allows for the reduction of the free energy barrier.

The heterogeneous nucleation rate has been plotted in Fig. 1.4 using the same data as previously, but for a nucleus cap, where $\theta = 15.6°$. It shows that the onset of nucleation effectively now begins around $\Delta T = 7\,\mathrm{K}$ undercooling, similar to common experimental observations.

The change of nucleation rate with undercooling for this situation is even more dramatic than in homogeneous nucleation as seen in Fig. 1.4, indicating that a 1 K change can produce a change in nucleation of four orders of magnitude.

1.2.3 Grain Refinement

The potency of some nucleant particles in promoting large numbers of nuclei during solidification is exploited industrially for grain refinement in castings. It has been shown recently (see review by Greer [5]) that grain refiners, such as Al–5wt% Ti–1wt% B added to aluminum melts, may result in a dramatic reduction in the undercooling to a maximum of 0.2 K before the onset of solidification. This indicates very potent catalysis for the nucleation of α–Al on the nucleant surface having a very low contact angle, in other words forming effectively a flat plate on its surface. Using an Al-based glass containing Al–5Ti–1B as a refiner, it has been possible to investigate the nucleation of solid α–Al on the nucleant or inoculant particles using transmission electron microscopy, since the process is slowed down due to the sluggish diffusion of aluminum atoms in the glassy matrix. This work reveals clearly that the aluminum solid nucleates only on the {0001} basal faces of TiB_2 particles, coated with a thin layer of Al_3 Ti, which is the actual nucleation substrate and may be affected by the matrix chemistry.

However, only 1% of the TiB_2 particles in the system produce Al grains in spite of possessing α–Al nuclei on their surfaces and the question arises concerning the inefficiency of grain production. Maxwell and Hellawell [6] assumed this occurred because the latent heat released by the initial nuclei raised the melt temperature causing the nucleation rate to fall. However, Greer has suggested a further barrier exists to the growth of the nuclei already formed, illustrated in Fig. 1.6. This involves the size of the inoculant particle and the surface area available for the nucleus growth. For a small particle, once the nucleus has been covered its surface, further growth of the nucleus cap may require a decrease in its radius (see Fig. 1.6(2)) and further undercooling is needed to achieve this. Eventually, a minimum radius r^* is achieved for a hemispherical nucleus ($2r^* = d$) at an undercooling for free growth, given by $\Delta T_{\mathrm{fg}} = 4\sigma/\Delta S d$, where ΔS is the entropy of fusion per unit volume and d is the nucleant particle diameter. At this undercooling, the nucleus will grow without further restraint (free growth) as its radius then increases with size to form a new grain. This is illustrated in Fig. 1.6, where for particles of size less than 6 μm growth is impeded after nucleation until undercooling has increased

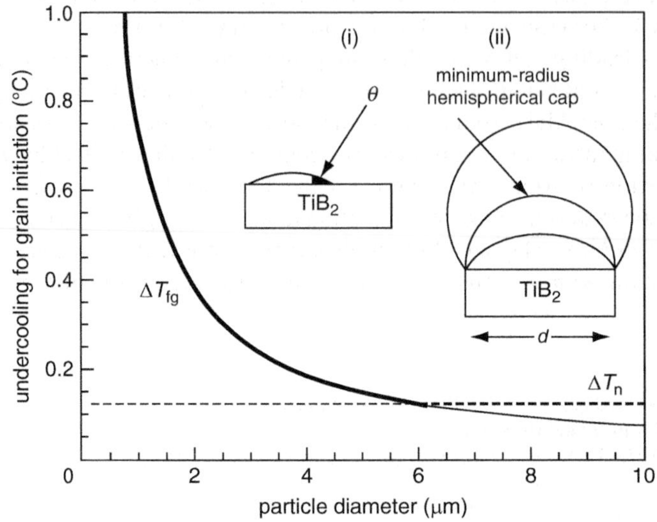

Fig. 1.6 The *bold line* indicates the undercooling necessary for grain initiation. The free-growth undercooling is calculated; the nucleation undercooling is schematic only. *Inset* (1) shows the classical spherical-cap model for heterogeneous nucleation. *Inset* (2) shows a cap of α–Al growing on an inoculant particle through the critical hemispherical condition [5]

to ΔT_{fg} to permit free growth; this is a region of free growth control. However, for large nucleant particles over 6 μm, growth of nuclei will occur spontaneously after nucleation (i.e., nucleation controlled).

Greer has checked the proposition that the free growth requirement may frequently control grain density under certain conditions by measuring the number of grains formed in a small melt, where isothermal conditions may be assumed during cooling controlled by constant heat loss from the system. The number of grains formed was measured as a function of the number of inoculant particles introduced into the melt. Predictions of the number of grains were made firstly on the basis of heterogeneous nucleation control and the effect of undercooling alone, as proposed by Maxwell and Hellawell [6]. This shows a linear increase in grain density with number of nucleant particles, (Fig. 1.7, region (A)), since at low particle concentrations, there is time for all particles to act as a nucleation source. However, nucleation is eventually unable to cope with high levels of nucleant particles and the number of grains formed falls off to form a plateau (region (B)). The second prediction assumes the criterion of free growth and uses the measured distribution of inoculant size in the melt, and shows the calculated number of grains does not fall off with the addition of particles but continues to rise providing good agreement with measured grain densities as shown in the Fig. 1.7.

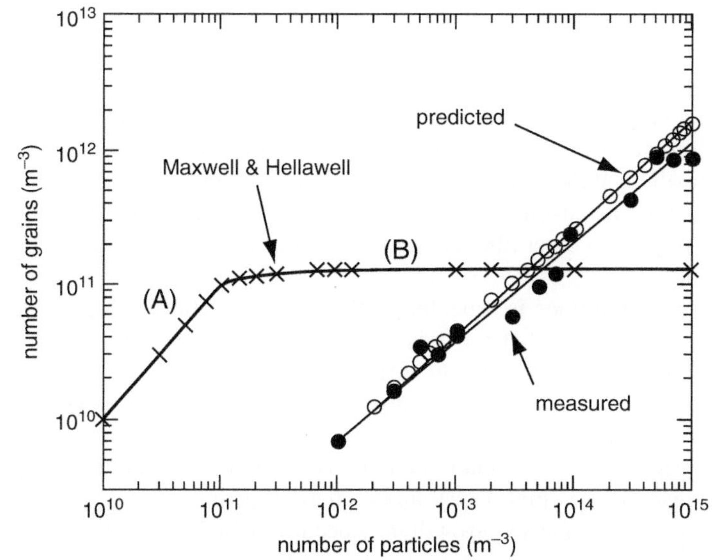

Fig. 1.7 The number of grains per unit volume as a function of the number of refiner particles per unit volume, showing a general trend to lower efficiency at higher addition level. Data from grain diameters measured in TP-1 tests (*closed circles*) are compared with predictions of the free-growth model (*open circles*) [5]. The predictions are qualitatively different from those of Maxwell and Hellawell [6] and are a much better fit to the data [5]

Another way of producing a fine grain size during solidification without introducing foreign particles, which may be deleterious to mechanical properties, is through grain multiplication as a result of the breakup of dendrites. This process will be discussed later (Sects. 1.5, 9.6, and 9.7).

1.3 Solid–Liquid Interface Structure

Some crystals, especially those in which directional bonding exists or possess complex structures, grow from the liquid phase with clear faceted surfaces, i.e., bounded by simple crystallographic planes. It is believed that these crystals have atomically smooth surfaces and growth proceeds either by the nucleation of two dimensional islands on such surfaces and subsequent growth by atomic attachment to the edges of this layer, or more likely by the growth at a ledge on the crystal surface produced by a crystal defect (e.g., a screw dislocation), which cannot be eliminated by the process.

In general, most pure metals and alloys possessing simple crystallographic structures do not grow with faceted interfaces, and these interfaces are considered to be diffuse, that is, the atoms at the interface are not clearly within either the solid or the liquid phases, but over a region, say of three atoms thick, move from one state to the other [2]. At this type of diffuse or rough interface, every atom in the region can migrate from one phase to the other and there are no special sites at the interface, such as atomic ledges in faceted growth. The consequence is that the interface moves forward continuously and uniformly at all points along it, with essentially a constant velocity in all crystallographic directions. To a first approximation, these structureless interfaces must also possess little anisotropy of surface energy, although some must still remain to determine the dendritic structure that evolves in alloy crystals during growth.

The type of interface is reflected in the entropy of fusion ΔS_f of the material [3]. It has been shown that for those materials with values of $\Delta S_f/R$ less than 1 (which includes all metals), that the interface between solid and liquid are diffuse (or rough), and give rise to non-faceted growth. Also the kinetic undercooling required for normal growth conditions is extremely small ($\sim$0.001 K), and for most growth calculations may be safely ignored compared with heat and solute diffusion. However, where $\Delta S_f/R$ is high (>2), as in silicon precipitating from Al–Si alloys or graphite from Fe–C, faceted growth may occur in which relatively high undercoolings are required for appreciable growth, and highly anisotropic growth rates and surface energies will exist.

1.4 Morphological Instability

In the last section, it was demonstrated that at the interface of a growing metal crystal, the kinetic undercooling is negligible and growth is essentially controlled by the transfer of heat or solute, to or from the moving interface. Local equilibrium is then said to exist at the interface. Under these conditions, however, the moving interface may become morphologically unstable: that is small perturbations formed at the interface, arising possibly from local disturbances in heat flow, can in certain circumstances become amplified, eventually developing a cellular or dendritic structure. In the case of pure metals, this can only occur when isolated free crystals grow into an undercooled melt (equiaxed growth); in columnar growth of crystals from the ingot surface, however, where a positive temperature exists into the melt, the interface is always stable to accidental growth fluctuations, since perturbation tips would move into hotter liquid and will melt back to restore a planar interface.

However in alloys, a different situation prevails for a moving planar interface. In this case, where the partition coefficient $k < 1$, solute is rejected by the solid into the liquid, forming a concentration profile as in Fig. 1.8a. Eventually, a steady-state solute distribution is established in the liquid as the rate of solute rejection is balanced by the diffusive flow below the composition gradient into the bulk liquid. It may be shown [3] that at this stage, the solute gradient at the interface $G_c = -(V/D)\,\Delta C_o$, where V is the growth velocity, D the diffusion coefficient in the liquid, and ΔC_o the difference between the interface and bulk solute concentrations. The interface concentration in the liquid (C_o/k) will possess a liquidus

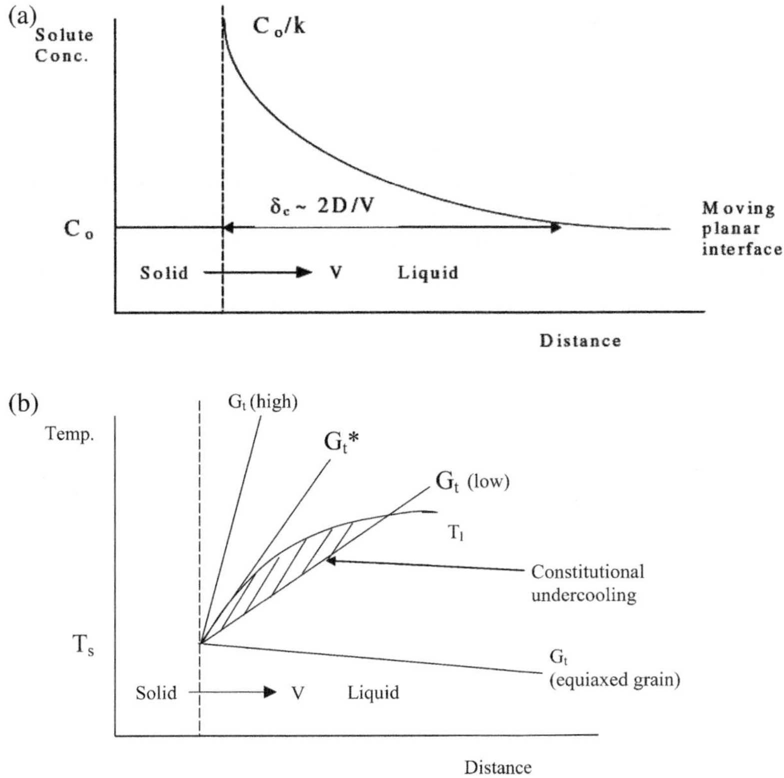

Fig. 1.8 Distribution of (a) solute and (b) temperature ahead of a moving planar interface showing the development of Constitutional Undercooling.

temperature, according to the phase diagram, below the liquidus temperature for the bulk of the liquid that is still at the original alloy composition C_o (see Fig. 1.8a). A liquidus temperature (T_l) may be constructed in the region ahead of the moving interface, as in Fig. 1.8b and the gradient of T_l at the liquid interface may be obtained by multiplying the solute gradient G_c above by m, the liquidus slope in the phase diagram. Where a steep temperature gradient actually exists within the liquid, G_t (high), it is clear that no undercooling exists ahead of the interface to encourage solid to grow. A low temperature gradient, G_t (low), however, produces a region that is metastable with respect to the formation of solid, and this is termed "constitutional undercooling." We might expect that for a surface perturbation in which solid enters this constitutionally undercooled region, it will be encouraged to grow and therefore, the planar interface is unstable. The critical condition will then occur when the liquidus temperature gradient equals the actual temperature gradient: that is $G_t^* = mG_l = -\frac{mV}{D}\Delta C_o$, and instability should occur when G_t is below this critical value. It should be noted that where equiaxed grains are growing within a melt, the temperature gradient must be negative (the latent heat must flow away into the surrounding liquid). As a consequence, an enhanced and extended region of undercooled liquid exists around each equiaxed grain in which dendritic growth is favored.

This approach considers only the thermodynamic restraint on interface instability and neglects the effect of surface tension, which might be expected to further constrain the development of instability, and furthermore the treatment provides no estimate of the scale of the perturbation. Applying a small sine wave perturbation to the interface, which is at local equilibrium but contains a surface tension modification, and examining whether the perturbation grows or shrinks, allows one to examine the effect of surface tension on stability. It transpires [3] that there exists a critical wavelength, $\lambda_c = 2\pi\,(\Gamma/\varphi)^{1/2}$, where Γ is the Gibbs–Thomson coefficient and φ is the degree of undercooling, below which the perturbation disappears and above which it grows increasingly rapidly. At the thermodynamic limit to instability, the perturbation wavelength approaches infinity and therefore, the planar surface is effectively stabilized by surface tension. A more detailed consideration of interface stability is given in [3].

1.5 Grain Multiplication

The generation of fine equiaxed grains by stirring during solidification in order to develop a thixoformable microstructure has much in common with the formation of equiaxed structures in conventional chill castings.

In chill castings, an outer columnar structure is formed at the mould wall of elongated dendrites growing into the melt with a specific crystallographic direction parallel to the direction of heat flow. Equiaxed dendrites may then develop in the ingot center with random orientation, growing approximately uniformly in all directions. It is known that this region of equiaxed grains is particularly sensitive to the superheat of the melt before pouring, and to the flow in the melt after pouring: high superheat reduces the volume of the equiaxed region, and flow tends to increase it. It has been shown that to prevent fluid flow by applying a strong magnetic field also reduces the amount of equiaxed structure.

Three possible mechanisms have been put forward to explain the presence of the equiaxed grains (see review by Flood and Hunt [7]):

(a) Constitutional undercooling ahead of the columnar grains, resulting in heterogeneous nucleation
(b) The so-called Big Bang theory, and
(c) Dendrite arm detachment

The possibility of heterogeneous nucleation occurring in the constitutionally undercooled region ahead of the columnar grains (which must be small given that the solute can diffuse very efficiently sideways from the growing dendrite tips) has been ruled out for two reasons. It has been shown by Chalmers [8] by imposing a mechanical barrier in the form of a thin metal cylinder between the inner and outer regions in the casting, which prevented the transportation of solid fragments from the outer columnar region to the center but did not affect the heat flow or the development of a constitutionally undercooled region ahead of the growing tips (see Fig. 1.8(a)), that the central equiaxed region then contained much fewer but coarser equiaxed grains compared to the identical experiment carried out without the barrier. The other piece of evidence against this nucleation theory is that experiments carried out containing alloy free of foreign nucleants still develop equiaxed structures. Although it is felt that constitutional undercooling cannot explain the origin of equiaxed grains (except where potent inoculants have been purposely added), nevertheless it must still play a part in their growth.

The Big Bang theory proposes that during pouring of the liquid alloy into the chill mould, crystals nucleate on the mould surface and are then washed into the center of the casting, some of which survive to grow into the equiaxed region at a later stage of solidification. The dendrite arm detachment concept is that remelting at the roots of arms on the columnar dendrites growing from the mould walls, resulting from local recalescence or thermal fluctuations from fluid flow, or from the forces created by flow, causes the arms to become detached from the mother dendrite crystal and carried into the bulk liquid. Essentially these two are similar mechanisms, distinguished only by time and place at which they occur. The Big Bang takes place during the pouring stage and is completed on the filling of the mould, whereas the dendrite remelting mechanism continues throughout the solidification process as long as circulation of the fluid by convection occurs. But both involve grain multiplication by the melting off (or detachment) of dendrite arms and their transport into the bulk of the melt. Once transported to the ingot center, growth of these free crystals occurs because the melt is constitutionally undercooled, and this eventually limits the growth of the columnar grains from the mould walls. It is thought that both these mechanisms may operate in the same system, perhaps at different stages in solidification. It is clear that the effect of superheat in reducing equiaxed grain formation is readily explained by the Big Bang, since most of the free grains formed at this stage will be remelted. However, continued stirring of the melt, after the superheat has dissipated, will assist dendrite arm detachment and encourage equiaxed grain growth.

Both these mechanisms are active in the recent slurry generation processes: the new rheocasting (NRC) process of Ube in which liquid alloy is poured into stainless steel cups and subjected to controlled cooling; and the semisolid rheocasting (SSR®) developed at MIT and now under license to Idra-Prince in which a spinning cold finger is introduced into the alloy melt to generate large amounts of small solid particles by dendrite fragmentation (see Chap 4, and Chap Sects 10.61 and 10.62).

1.6 The Growth and Spheroidization of Equiaxed Grains

It has often been claimed that the near spheroidal or rosette morphology of grains growing within the stirred melt arises from their circulation in the semisolid region, passing through hot and cold zones, alternately melting and growing, which results in degenerate dendrites when solidification is complete. However, it is also possible that such morphologies might develop under steady growth conditions.

It has been stated already that under the growth conditions experienced by equiaxed grains in alloys, constitutional undercooling will exist around them leading to dendritic growth. It might be supposed, therefore, that convective flow across the solid–liquid interface produced by stirring would help to remove the solute build-up around them, reducing the constitutional undercooling, and so might restore planar (or spherical) stability. However, the theoretical investigations by Vogel and Cantor [9, 10] into this problem of growth in alloy systems under laminar flow conditions, by modeling heat and solute diffusion away from a perturbed spherical particle using boundary layer approximations, demonstrated that not only did the growth rate of the interface increased (resulting from increased heat transfer), but the interface became more unstable, breaking down into a finer dendritic structure. However, these conclusions are based on the assumption of laminar flow in the liquid and recently, Qin and Fan [11] have carried out similar theoretical computations but including turbulent flow. Their results support Vogel and Cantor for laminar flow but show that under turbulent conditions flow is able to penetrate inside the perturbed interface and remove the accumulated solute. This can lead then to restabilizing of the interface, first under less intense flow to generate a rosette morphology and eventually at greater turbulence to promote the rapid growth of spherical particles. It is claimed that such behavior is observed in "twin-screw rheomolding" [12].

Chapter 2
Characterization of Microstructure in Semisolid Slurries

It has been understood for some time that the flow behavior of semisolid metal slurries, and the properties of parts shaped from the slurries, depend both on the fraction solid and on the size, distribution, and morphology of the solid particles within the liquid matrix. The fraction solid f_s is essentially determined by the temperature for a given alloy; the other features, however, are a function of the history of preparation and it is known that slurry flow is enhanced by achieving fine spherical particles.

2.1 Fraction Solid

Fraction solid (f_s) can be determined from phase diagrams for simple alloy systems if equilibrium may be assumed, or from direct measurement on rapidly quenched specimens, using metallographic techniques on sectioned surfaces, such as line intercept: $f_s = L_a$, where L_a is the total intercept length in the a particle phase per unit length of test line. Likewise, the average particle size (assuming spheres of constant diameter d) may be obtained by the measurement: $d = L_a/N_a$, where N_a is the number of grains per unit area of sectioned surface. It should be noted that the fraction solid obtained from phase diagrams is strictly a weight fraction and only coincides with the volume fraction obtained from metallographic measurements above when the densities of liquid and solid are considered equal and shrinkage is therefore ignored. Other techniques have been used to obtain these quantities, but are indirect, requiring calibration, and are probably less reliable. However, see in situ X-ray microtomography in Sect. 3.4, where the measurement is carried out in the semisolid condition to avoid quenching artifacts.

2.2 The Shape Factor

Spheroidized particles within a semisolid alloy slurry are an essential prerequisite structure for effective thixoforming, and it is necessary therefore to develop a quantitative description of this property. One commonly used descriptor, which may be obtained from two-dimensional sections through a complex morphological structure, is to measure the perimeter length P of an object particle present from a

David H. Kirkwood et al., *Semi-solid Processing of Alloys*,
Springer Series in Materials Science,
DOI 10.1007/978-3-642-00706-4_2,

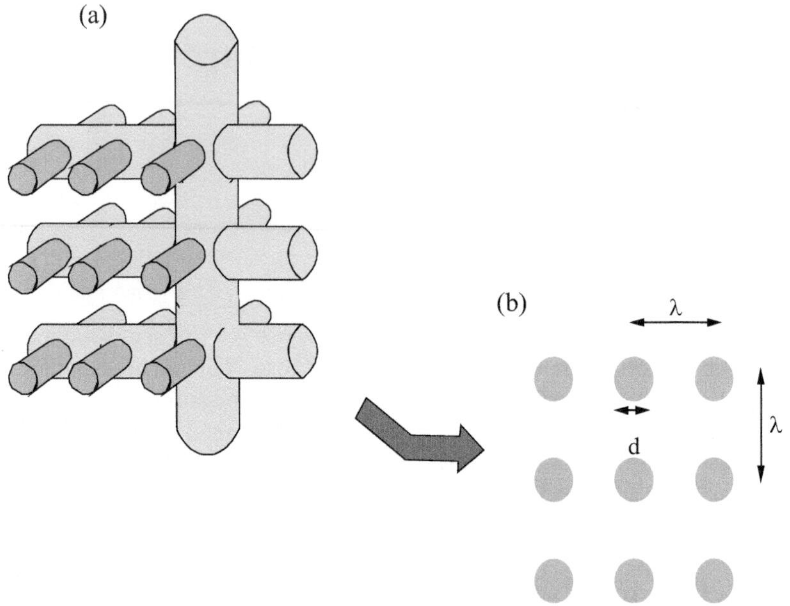

Fig. 2.1 Schematic representation of complex Dendrite Structure, (a) 3D view (b) Cross Section.

section, and its sectioned surface area A. The dimensionless shape (or form) factor is then defined as: $F_o = P^2/4\pi A$ (note that some authors use the reciprocal of this factor); in the case of a section, which appears as a circle, $F_o = (2\pi r)^2/4\pi \cdot \pi r^2 = 1$, and convoluted perimeters would clearly produce values in excess of unity. However, it is clear from the diagram in Fig. 2.1 that circular sections may not only arise from spheres, but also from much more complex morphologies in three dimensions, such as sections through cylindrical dendrite arms, and such complex morphologies are not reflected in this simple shape factor.

To overcome this drawback, Loué and Suery [13] have introduced a different dimensionless shape factor defined as: $F_g = 1/(6\pi f_s)\left(S_v^2/N_a\right)$, where S_v is the surface area per unit volume of the particulate phase and N_a the number of grains per unit area of sectioned surface. Both of these values are in principle able to be measured by direct metallographic techniques from two-dimensional sections. For the situation of a monodispersate of spheres having a uniform diameter d, we have:

$$S_v = \pi d^2 N_v = \pi d^2 \cdot 6f_s/\pi d^3 = 6f_s/d$$

and

$$N_a = N_v d = \left(6f_s/\pi d^3\right) d = 6f_s/\pi d^2$$

so that

$$F_g = \frac{1}{6\pi f_s}\left(\frac{6f_s}{d}\right)^2\left(\frac{\pi d^2}{6f_s}\right) = 1.$$

Some insight into the physical meaning of this shape factor may be gained from the following simple consideration. Imagine a simple square array of cylindrical dendrite side arms, all belonging to the same grain. If each arm has a diameter, d, and the arms are separated by a distance λ (see Fig. 2.1), then $f_s = \pi d^2/\lambda^2$, the interface area per unit volume $S_v = \pi d/\lambda^2$ (that is inversely proportional to λ for a given f_s) and the number of grains per sectioned surface area $N_a = 1/(n\lambda)^2$, where $n\lambda$ is approximately the grain size.

Hence $F_g = \frac{1}{6\pi f_s}\frac{S_v^2}{N_a} = \frac{1}{6\pi f_s}\left(\frac{\pi d}{\lambda^2}\right)^2(n\lambda)^2 = \frac{2}{3}n^2$, so that this shape factor is simply related to the number of side arms possessed by a dendrite. For $n = 10$ side arms, $F_g = 67$, which is similar to the value observed in a conventionally DC cast billet, see Fig. 3.1a. An anomaly exists for a single independent cylindrical rod, where $n = 1$, giving $F_g = 0.67$, smaller than the value for a sphere; however, a rod can hardly describe a dendritic array and appears the consequence of an over-simplified model not intended to predict the complete morphological change of a dendrite with time.

Any dispersed system will tend to minimize its surface free energy and therefore, its surface area with time. This requires that in complex convoluted morphologies with high surface energy per unit volume (S_v), the particles will first spheroidize (i.e., $F_g \to 1$), followed by the process in which the smaller particles or grains dissolve to feed the larger ones (akin to Ostwald ripening). The first stage will proceed more rapidly because of the shorter diffusion distances involved, and the objective of the heating stage for thixoforming is to achieve this first stage of spheroidization without entering the second stage of grain growth.

2.3 Contiguity and Continuity

Gurland [14] has introduced the concept of contiguity ratio defined as the average fraction of the surface area of a particle shared with neighboring particles of the same phase:

$$C = \frac{2S_{\alpha\alpha}}{2S_{\alpha\alpha} + S_{\alpha\beta}},$$

where $S_{\alpha\alpha}$ is the area of interface between α particles, $S_{\alpha\beta}$ between α particles and the β matrix and the factor 2 appears because for contiguous particles the interface are counted only once. For completely isolated particles, $C = 0$, whereas for grains in total contact (i.e., $f_s = 1$), $C = 1$. This parameter is not a function of the shape or size of the particle; it is concerned only with contacts between grains and their immediate neighbors and may be regarded as a measure of *short-range* continuity as experienced in particle clustering. It is expected to depend on both processing and subsequent heat treatment.

Table I. Experimental Measurements on Silver-Bakelite Aggregates[9]

Composition, vol pct Ag	Mean Number of Contacts per Silver Particle	Electrical Resistivity, ohm-cm
10	0.40	3.5×10^{11}
20	1.00	2.5×10^{11}
30	1.25	5.0×10^{10}
36	1.33	1.2×10^{6}
37	1.52	10
38	1.45	9.2×10^{-2}
40	2.47	3.1×10^{-2}
50	2.74	2.9×10^{-2}

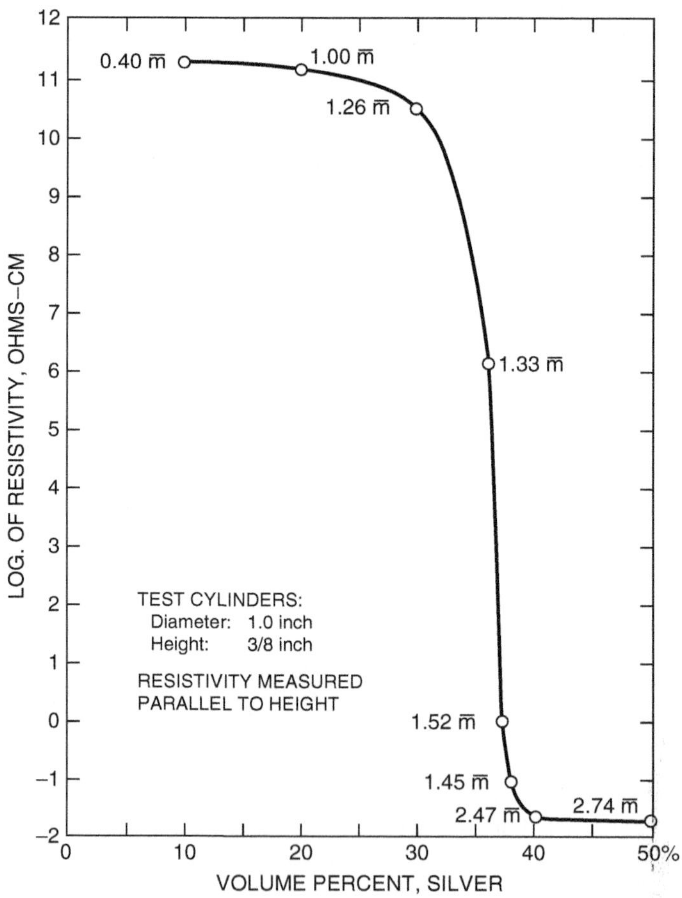

Fig. 2.2

Gurland [15] has used a different measure to describe the degree of *long-range* continuity of a dispersed phase within a matrix, defined as the probability of infinitely long chains of connected particles being formed, which are orientated randomly and cross-linked in 3D, to provide a rigid skeletal framework. This is based on the average number of contacts ($\bar{m}$) that a particle makes with neighboring particles of the same phase. Experimental measurements on compacts of silver spherical particles embedded in bakelite show that the resistivity falls off dramatically as the volume fraction of silver increases from 0.36 to 0.38 (Fig.2.2). This is interpreted to indicate that long chain structures develop rapidly over this compositional range. The average number of contacts of a silver particle were measured by quantitative microscopy on specimens in this region, and it was established that long range continuity occurred when $\bar{m}^* > 1.5$. This critical value for m is supported by two simple models investigated, a branching chain model and a percolation model, both of which also demonstrate the dramatic transition to long range chain structures with composition. In the light of the above, it would appear that for the conditions under which semisolid forming is carried out at $f_s = 0.5$–0.6, long-range skeletal structures may be expected to be created, thus contributing to the observed stiffness in these materials.

In a later paper, Lee and Gurland [16] introduced the concept of "continuous volume" to explain the hardness of cemented carbides. They showed that the yield strength σ of a composite material containing a skeletal structure of particles in a continuous matrix could be given by:

$$\sigma = \sigma_p V_c + \sigma_m (1 - V_c),$$

where σ_p and σ_m are the yield stresses for the particulate and matrix phases respectively, and V_c is the fraction of continuous volume of the particulate phase bearing the load shown to be equal to CV, where C is the contiguity ratio and V the volume fraction of particulate. This expression appears to describe the hardness of composites containing tungsten carbide particles in a cobalt matrix very well. If we use it for semi-solid slurries, $\sigma_m = 0$, since the liquid matrix will bear no load, and writing f_s for V, we have $\sigma = \sigma_p C_f$ where C_f is the continuous volume fraction of the solid skeleton. Inserting a value of $\sigma_p \sim 2\,\text{MPa}$ (obtained from extrapolating high temperature data for pure aluminum [17] to 570°C), and using experimental measurements of $C_f = 0.15$ [18], we obtain $\sigma \sim 0.3\,\text{MPa}$, which is more than sufficient to prevent sagging of a slug of 10 cm height under gravitational force (~3 kPa).

Chapter 3
Evolution of Microstructure in Semisolid Alloys During Isothermal Holding (Soaking)

3.1 Introduction

In a recent review of coarsening, Flemings [19] states that this process refers to the growth "of solid regions of low curvature at the expense of regions of higher curvature," and this includes "the growth of larger particles or dendrite arms with the simultaneous dissolution of smaller particles or arms (so-called 'ripening'), the filling of spaces between particles or dendrite arms ('coalescence') and the break-up of dendrites ('dendrite multiplication')." This is a view adopted in the present chapter. The driving force is of course always the reduction of total solid–liquid interface area and the reduction in the associated interfacial energy, and the general mechanism by which this is achieved is diffusion of solute atoms through the liquid from concentration gradients established between regions of high and low curvature. Owing to the complex geometries that may exist in coarsening semisolid systems with competing fluxes between different particles, which can change with both time and place, simple kinetic equations to describe the overall coarsening process is probably not possible. However, where the geometry of the solid may be clearly described, for instance in terms of dendrite arms formed early in solidification, simple kinetics can be derived. More complex geometries may develop later during coarsening that are more difficult to provide with an adequate geometrical description.

It is convenient (and to some extent accurate) to distinguish further between two coarsening processes occurring during the isothermal soaking of dendritic or near dendritic microstructures: spheroidization and ripening. The former precedes ripening, and involves the melting back or thinning of dendrite arms, and in some cases, their melting off [7], as a result of high surface curvatures and short diffusion distances providing high solute gradients in the liquid around the grain. The negative surface curvature at the dendrite root is in contact with high solute in the liquid, whereas at the tip the liquid is poor in solute, giving rise to these gradients. Consider a fine dendritic particle of 100 μm size, coarsening isothermally by the arms melting back from their tips to form eventually a spheroid or globule of approximately similar diameter. When the arm spacing λ is of the same order as the particle size, the dendritic structure has effectively disappeared and replaced by a globular

David H. Kirkwood et al., *Semi-solid Processing of Alloys*,
Springer Series in Materials Science,
DOI 10.1007/978-3-642-00706-4_3, © Springer-Verlag Berlin Heidelberg 2010

particle. This will occur for $\lambda = 100\,\mu$m in a time $t_0 = mHC(1-k)\lambda^3/128D\sigma T$ [20], where C is the solute concentration in the liquid, k the partition coefficient, H the heat of fusion, T the absolute temperature, σ the surface energy, and m the liquidus slope. Using data from the Al–4.5%Cu system [20], t_0 is determined to be about 6 min. Although this is a rough calculation, it does indicate that in the period of ingot soaking in the semisolid state, there is time for fine dendrite structure on the scale above to develop spheroidal shapes. This stage of coarsening only involves *intragranular* diffusion (i.e., in the liquid within the dendrite) and the grain density remains essentially constant.

Once the dendrite structure has effectively disappeared (but before spheroidization is fully completed), the second process of Ostwald ripening can begin, which involves the diffusion of solute in the liquid *between* grains (*intergranular* diffusion). This is a slower process because interface curvatures are less than in dendritic structures, and diffusion distances are greater, both leading to lower solute fluxes. The classical analysis of this situation for widely separated spherical particles in a stagnant matrix is due to Lifshitz and Slyzov [21] and Wagner [22] (LSW theory), and predicts cubic behavior with time: $d^3 - d_0^3 = k_2 t$, where d is the particle diameter at time t, and d_0 is the initial diameter. Using the value of k_2 determined by Loué and Suéry [13], and $d_0 = 100\,\mu$m, the particle diameter after 1 h is $115\,\mu$m, which may be regarded as negligible growth for thixoforming purposes. However, this neglects coarsening by coalescence of particles by collision and joining together, which may be important in some situations (see Sect. 3.4).

3.2 Coarsening in Al–Cu Alloys

Most of the early experimental work in this field of coarsening was carried out on Al/Cu alloys much by Flemings et al. [23] following their original work on solidification structures. Kattamis et al. [24, 25] have more recently added to this literature, again investigating the changes in morphology of directionally solidified dendritic structures during solidification, as well as examining changes during isothermal treatment of sections. It was observed that under isothermal conditions, the dendrite structure fragmented after about 30 min, thereafter forming a spheroidal particle structure which subsequently coarsened very slowly. Using standard optical metallography techniques, they obtained information on changes of secondary arm spacings (λ) and the total surface area per unit volume (S_v) as a function of time. Both parameters agreed satisfactorily with different geometrical models [20, 25] based on diffusion of the solute in the liquid from regions of high curvature to those of low, and demonstrated that copper diffusion was the rate-controlling step.

The work of Poirier et al. [26] using grain refined Al–15.6 wt% Cu alloy, had a fine equiaxed dendritic structure initially, whose secondary arm spacing rapidly increased with isothermal holding time until the grains developed an essentially globular shape after around 20 min, depending on the original fineness. The arm spacing increase followed a $t^{1/3}$ time law, again consistent with solute diffusion from the dendrite tips resulting in their melting (dissolving) back. Subsequent coarsening of the globular particles was again slow with $S_v \propto t^{-1/n}$, where $n > 5$. This is clearly much slower than predicted by the LSW theory of ripening, which assumes growth of separated particles by diffusion through the matrix and it was felt to imply coalescence in which particles join together.

3.3 Coarsening in Al–Si Alloys

The much of the work carried out to investigate the change in microstructure during isothermal holding has been directed to the Al–7Si alloys, A356, and 357. These are the alloys most commonly used in commercial thixoforming that produce around $f_s \sim 0.5$ on melting the Al–Si eutectic at 577°C. It is clearly of great practical importance to understand the conditions in which ideal thixoforming structures may be formed, that is possessing fine spheroidal or globular solid particles within a liquid matrix.

A key investigation in this area is that of Loué and Suery [13], in which S_v, N_a, and the shape factor F_g were measured as a function of time at 580° C ($f_s = 0.45$) in alloy produced by DC casting, with and without electromagnetic stirring (MHD). The difference in initial microstructure and after isothermal holding for 5 min and 15 h is shown in Fig. 3.1, revealing a highly dendritic structure in the conventional unstirred casting compared to the degenerate structure in the MHD casting. The measurements of S_v and N_a are shown in Fig. 3.2a, b, respectively, for times up to1500 min (about 1 day), the former indicating surprisingly little difference between the two production routes. However, the grain density, N_a, falls dramatically in the MHD material, but remains effectively constant in the conventional casting. A further difference in behavior can be observed in the shape factor F_g (see Fig. 3.2c), where it is very clear that the MHD structure rapidly spheroidizes within 30 min, whereas the conventional casting still has a complex morphology even after 24 h. These results have been interpreted as indicating that the first process to occur during isothermal holding of semisolid structures is the disappearance of dendrite arms within each grain, either by melting back of short arms, or by the coalescence of neighboring arms of the same grain that necessarily have the same crystallographic orientation and may lead to the entrapment of liquid between longer dendrite arms. Clear evidence of both these processes were obtained in the early work of Chien and Kattamis [27] and particularly from the quenched steady state growth of Al–Cu alloys of Young and Kirkwood [28]. Both are diffusional processes driven by solute differences in the liquid originating from the curvature of the interface and lead eventually to spheroidization accompanied by removal of entrapped liquid and an overall reduction of surface area in the system. Only after a degree of spheroidization with the removal of entrapped liquid has been achieved, however, the process of grain coarsening can begin, in which the smaller grains

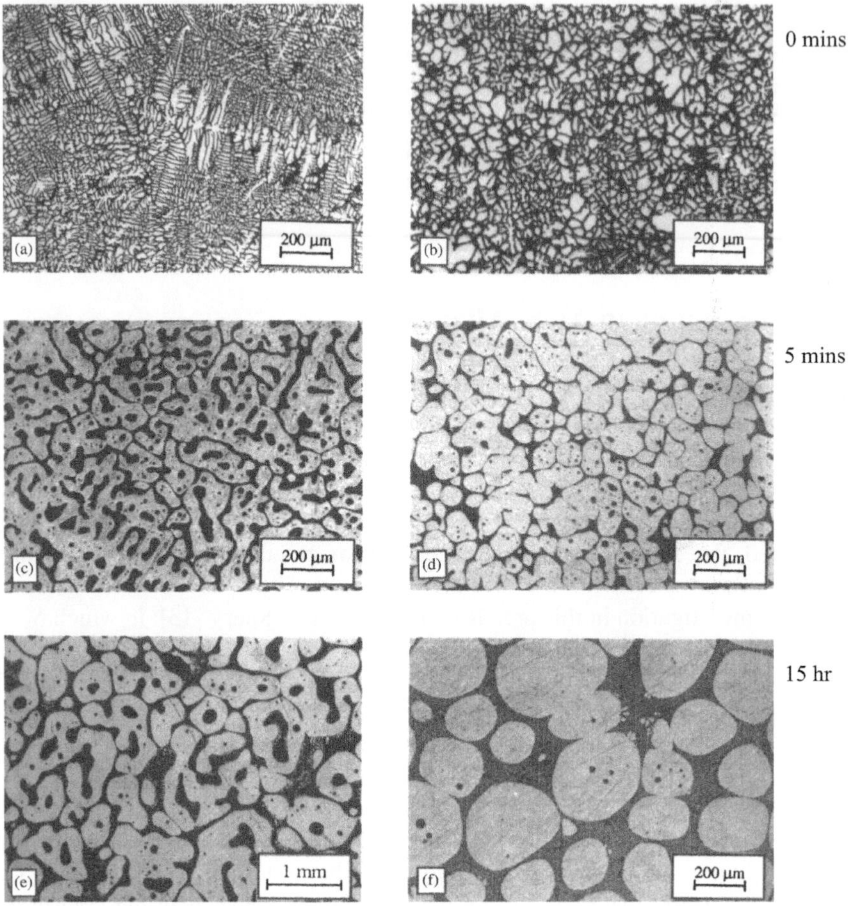

Fig. 3.1 (**a**, **c**, **e**) Conventional casting after 0 min, 5 min, and 15 h soaking, (**b**, **d**, **f**) MHD casting after 0 min, 5 min, and 15 h soaking

disappear as the larger ones grow by solute diffusion (Ostwald ripening) or by the coalescence of freely moving individual grains or agglomerates to form even larger agglomerates, leading to a fall in N_a. This is why the highly dendritic morphology of the conventionally cast structure maintains a constant grain density shown in Fig. 3.2b.

Tzimas and Zavaliangos [29] have also examined the microstructural changes in this alloy, as prepared by electromagnetic stirring, occurring during the first 5 min of isothermal treatment at $f_s = 0.45$, which covers the period of time normally used in industrial practice. They observed different initial microstructures within the cast billet, which had a highly dendritic columnar structure at the perimeter, whereas the center consisted of "rosette"-shaped particles. It is postulated that these rosettes

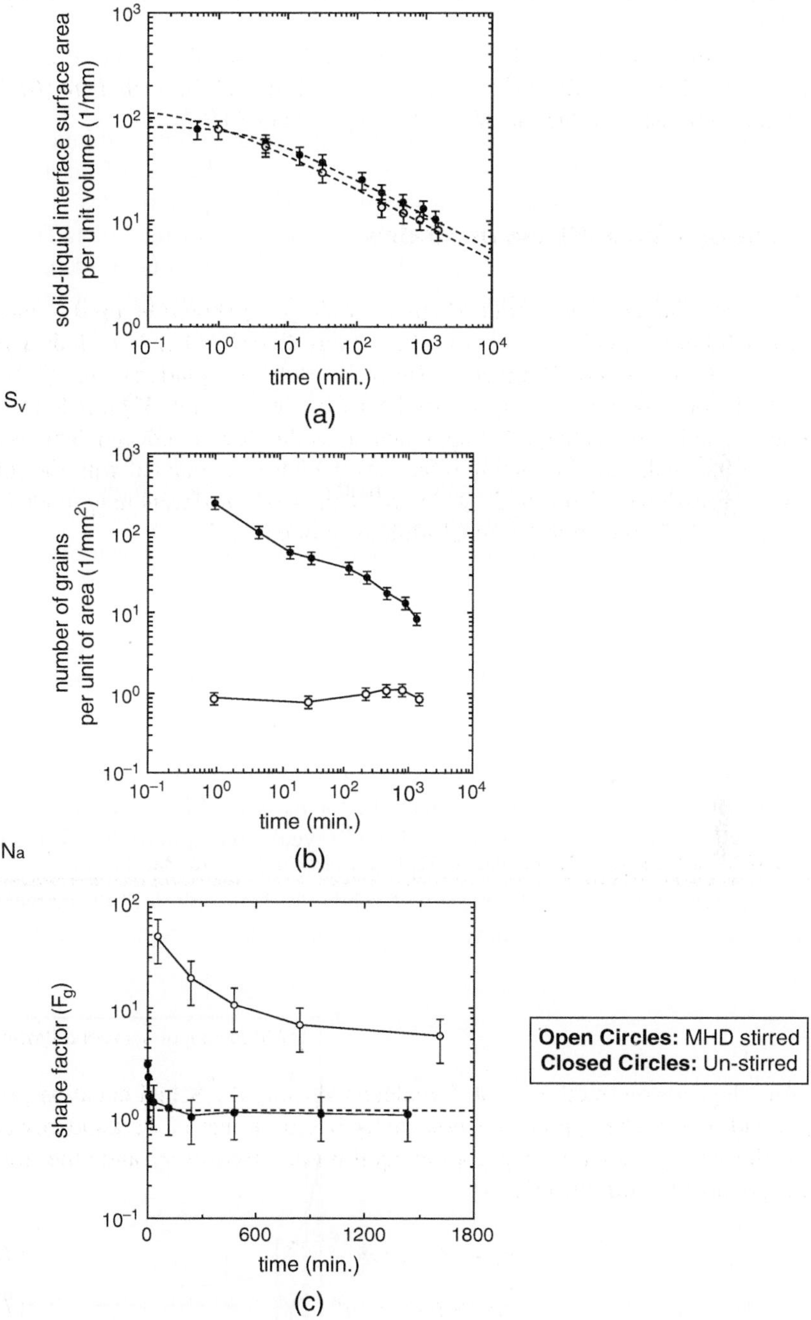

Fig. 3.2 Microstructural changes in DC cast Al/7%Si Alloys during isothermal heat treatment at 580 degC.

are created in the solidification process as a result of the circulation of the fluid within the ingot sump generated by the moving electromagnetic field that detaches dendrite arms from the growing perimeter region and circulates them into the hot sump center. Here, they partially melt and spheroidize initially to form rosettes, and in the process of circulating within the central sump, they collide and join together where the grain orientations are favorable to form agglomerates of rosettes having low energy boundaries separating the grains. These agglomerates or clusters are eventually deposited at the inward growing solid wall of the billet. The clusters of rosette-shaped grains can contribute to the rapid coarsening of grains during isothermal treatment in the semisolid state through the elimination of low angle boundaries by migration or rotation of the grains within the cluster. The shape factor (F_o) in both structural regions shows a rapid decrease in the first few minutes (as with Loué and Suery) so that little difference exists between perimeter and interior, in both shape and grain size. Allowing for the different measures of shape factor, the agreement on grain size and morphology between the two investigations is quite satisfactory: within the first few minutes of isothermal treatment, rapid spheroidization of the grains takes place with limited grain growth (from 70 to 100 μm). This is quite acceptable for thixoforming. In contrast to this, the shape factor in conventional unstirred DC billets in these short times (and even in much longer periods) is very high, and would not be suitable.

Other production routes to form appropriate semisolid slurries in this alloy system have been investigated by both these researchers. By cold working, followed by recrystallization and partial melting (SIMA route [30]), Loué and Suery [13] showed that cold work resulting from 25% reduction in section led to a reduction in grain size down to 55 μm, irrespective of the starting grain size; further cold work had no apparent effect. The evolution of the shape factor F_g of the SIMA material as compared to MHD alloy is shown in Fig. 3.3, and reveals that spheroidization is even more rapid, essentially being achieved in less than 5 min isothermal holding. This is attributed to the initial small grain size and nondendritic morphology of the partially melted microstructure. Furthermore, it was shown that the coarsening rate at 580°C

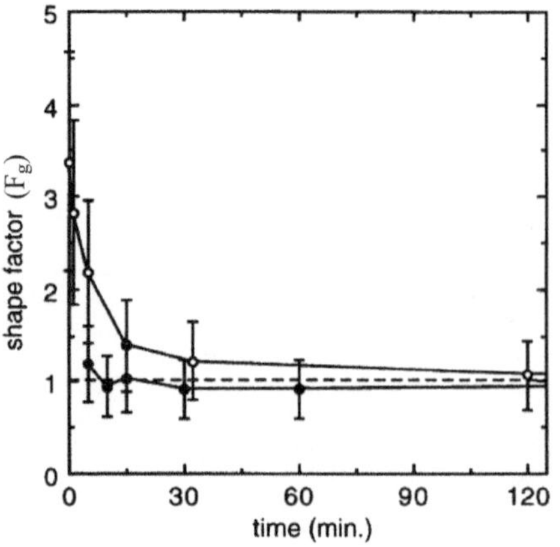

Fig. 3.3 Effect of Cold Work on particle spheroidisation (closed circles) compared to undeformed alloy (open circles).

of the SIMA alloy was lower than that of the MHD for these essentially spheroidal particles, and greater than expected from Ostwald ripening alone. It was therefore suggested that the coalescence of particles or aggregates must play a significant part in growth, and that the clustering of particles with low energy grain boundaries (see earlier section) as expected in MHD, might explain the higher coarsening rate seen in this material.

Similar findings to the above were reported by Tzimas and Zavaliangos [29] in alloy, which had been heavily warm worked before partial melting (RAP route), although this resulted in a larger grain size (90 μm). However, no grain growth could be observed during isothermal holding in the semisolid state over 5 min, nor was any shape factor change reported.

Finally, the effect of grain refinement and spray casting as fine grained source material for thixoforging was examined by these workers in view of the fact that small grains should spheroidize more rapidly. This was demonstrated using a grain refined alloy with Ti_5B addition to a permanent mould casting. The shape factor was significantly reduced after isothermal holding below that without refinement after 35 min, but still well above that for the MHD stirred material however. The spray cast alloy generates the finest grain size of all, 40 μm near-spherical particles in this alloy system, although since the diameter increases to 70 μm in 5 min, it is not clear that this is a great benefit in thixoforming given the inevitable increased cost of the source alloy.

A disadvantage in studying the Al–7Si alloy system is that 50% volume fraction liquid is formed immediately on melting the eutectic, which prevents studies on f_s of greater than 0.5. Using Al–4Cu alloy, Tzimas and Zavaliangos investigated the short time coarsening of spherical particles from both SIMA and spray-cast sources at $f_s = 0.6$ and $f_s = 0.9$. The results indicated that the spray-cast alloy coarsened more slowly and that, contrary to prediction, the high fraction solid slurry coarsened the slowest. This might be explained in terms of coarsening by particle coalescence involving the rotation of grains to form low energy boundaries and their migration. Where grains are initially random in orientation to one another, as in spray-casting, this is a slow process. Furthermore, where fine insoluble precipitates or gas pores exist within the solid, as noted in some spray-deposited alloy, this will further hinder brain boundary migration and the coarsening process.

3.4 X-ray Microtomography of Alloy Slurries

Microstructural changes occurring during the isothermal treatment of semisolid slurries in most investigations have involved quenching followed by sectioning. The problem with this procedure is, firstly, the quenching is never rapid enough to prevent further growth of the solid so affecting microstructural measurements, and, secondly, solids that are connected in 3D, may not be revealed as such in a 2D section. Serial sectioning [31] carried out to investigate this is limited by the fineness of sectioning that can be mechanically performed.

X-ray microtomography overcomes both these limitations by employing a high intensity X-ray beam generated using a synchrotron to penetrate through a small sample and recorded by a high speed digital camera [32]. Using this set-up, the microstructure can be continuously scanned (without quenching) during isothermal treatment, and a scan can be carried out every minute on the same small volume (during which negligible change is observed), in which the change of microstructure with time may be recorded with a spatial resolution of better than 3 μm.

Suery et al. [33] have used this technique at Grenoble to investigate microstructural changes during the partial melting of Al–15.8% Cu alloy, held at 555°C in the semisolid region for up to 80 min. Over this period, a number of global measurements were made on a specific volume within the sample: the number of particles per unit volume (N_v), the mean shape factor, the average coordination factor, the average particle volume (V) and effective radius (R), and the particle surface area per unit volume (S_v).

One surprising result was the coordination number of each particle (the continuity factor $\overline{m}$ in Sect. 2.3) with neighboring particles in an undisturbed sample, which revealed that there are no free unattached particles but most on average are connected to six or seven neighbors. This probably explains the initial apparent yield point in undisturbed slurries. Other significant results were:

1. The volume fraction solid with time, showing an increase in the first 30 min after which it levels out. This is attributed to diffusion in the solid leading eventually to complete equilibrium between solid and liquid, and the absorption of entrapped liquid within the particles. It was also shown that in quenched samples there was 20% greater solid than the nonquenched, revealing the possible large error introduced by quenching
2. The shape factor, determined from $6V\sqrt{(\pi/S^3)}$ where V is the volume and S the surface area of the particle, and is 1 for a sphere. The mean value rises rapidly from 0.66 to 0.82 in 25 min, and then slowly to 0.85, but complete spheroidization is not achieved

Since it is clear that the averaged values of N_v or S_v cannot establish the coarsening process unambiguously, attention was directed by these authors to measure the changes occurring locally between particular particles, and the growth or shrinkage of the necks between them. Two general extreme types of behavior were observed: (1) that between particles of very different sizes in which the smaller shrinks continuously with time in a process similar to Ostwald ripening, feeding larger neighbors (Fig. 3.4); and (2) that between particles of essentially equal size resulting in their coalescence eventually into a single particle (Fig. 3.5). In the latter case, the contact area of the neck grows continuously with time, whereas in the former it decreases. However, there are many intermediate situations in which the neck first grows then shrinks, or even remains static. These diverse behaviors are no doubt the consequence of the complex and changing particle environments, and make the description of coarsening in terms of simple global kinetic equations impossible. However, it would appear that the morphological changes are all driven by solute diffusion through the liquid matrix by gradients established by local surface

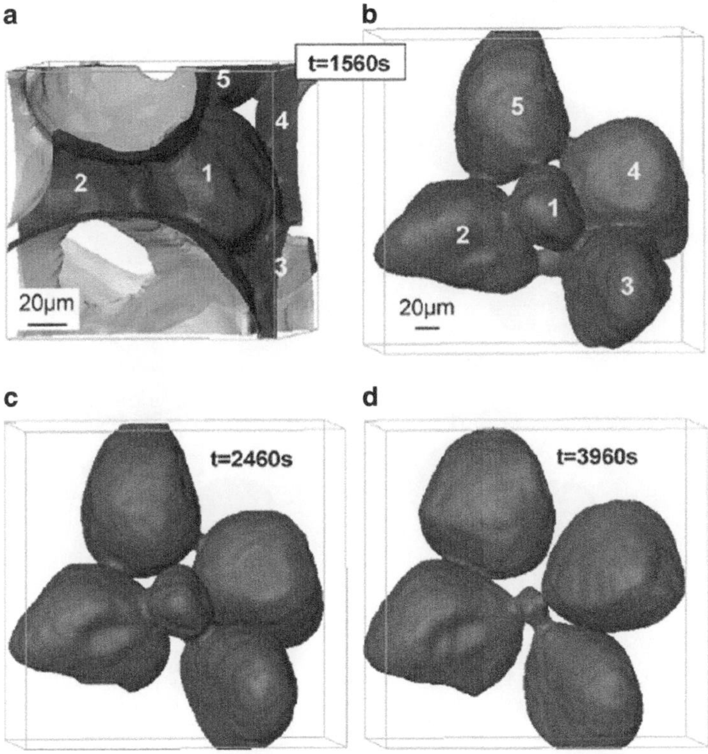

Fig. 3.4 Observations of neck representative of trend with unequal sized particles

curvature. Furthermore, the finer morphologies, such as dendrite arms and small particles will coarsen rapidly and disappear first because of their associated higher curvatures.

The plots of $1/N_v$ and V against time both indicate a change in kinetics in this work at around 50 min: The average particle volume V increased linearly with time up to this point (see Fig. 3.6) before falling off. The linear portion indicates a $t^{1/3}$ growth in radius that is consistent with ripening involving the absorption of small particles by larger neighbors by a process akin to Ostwald ripening as proposed by Courtney [34]. The particle density (N_v) changes as $1/t$ up to 50 min, which is also consistent with a ripening process, thereafter the rate of particle removal decreases presumably because the smaller particles have disappeared by absorption, and the dominant process is now the coalescence of essentially equally sized particles – a much slower process, because of lower solute diffusion gradients. This would account for the slower growth of the average particle volume V and radius R in Fig. 3.6 after 50 min. Also, a slower growth rate is obtained in fitting S_v as a function of time where the time exponent is around 1/7. The explanation for this low exponent is not entirely clear but it is suggested that the influence of neighboring particles, with complex and competing diffusion fields must be involved at this later stage in

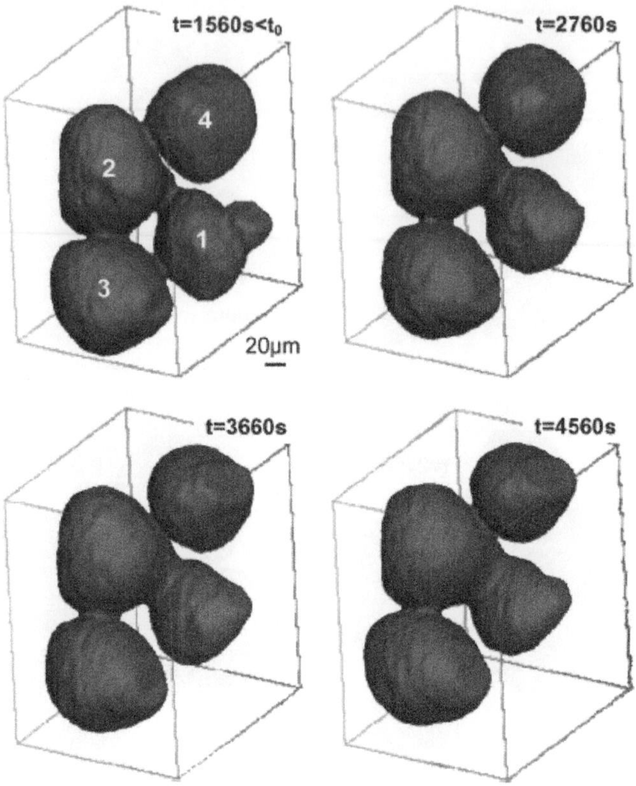

Fig. 3.5 Observations of neck representative of trend for equal sized particles

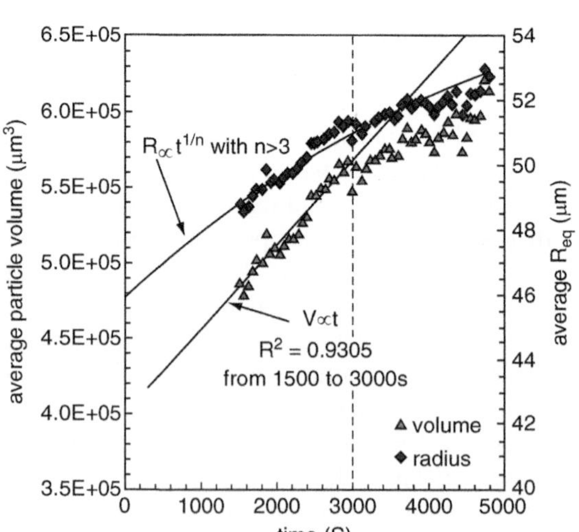

Fig. 3.6 Variation of the average solid particle volume and equivalent radius with isothermal holding time

coarsening. The rate of growth of the necks between particles was also examined by X-ray microtomography. It was found that the average neck radius increased with time as $t^{1/5.8}$. Also, a number of individual necks were measured and the average of the continuously growing necks (i.e., those exhibiting pure coalescence) increased as $t^{1/4.8}$. Both are close to the value, predicted by Courtney of $t^{1/5}$, based on diffusion of solute through the matrix to the neck formed between two particles of equal size.

A more detailed account of the results and interpretation of this work are given in [33].

Chapter 4
Recent Developments in Slurry Formation

Despite the proven improvement in the quality of products formed by semisolid processing, providing both better surface finish and mechanical properties after heat treatment, the main obstacle to progress in the industrial use of thixoforming are due to the additional costs involved in.

1. The production of *solid* billets having the correct internal structure, e.g., by the MHD process
2. The reheating, melting, and soaking of the billet to reform the semisolid structure
3. The scrap produced by each thixoformed article, which must be returned to the billet manufacturer for reconstitution as appropriate billets

These cost problems have led back in recent years to a reconsideration of the rheoforming route (the production of semisolid alloy slurry on site followed by immediate injection of slurry into the die), that was originally used at MIT, thus bypassing casting of solid by MHD.

The UBE process [35] has been developed whereby rapid cooling of the aluminum alloy in a cylindrical steel crucible forms a semisolid billet. The semisolid billet and crucible are then transferred to the die casting machine by robot arm, where the billet is dropped into the shot sleeve for immediate injection. The scrap from the operation is collected ready for remelting in the holding furnace on site for further casting. This route clearly obviates the need for MHD stirring to produce a nondendritic structure, and the heating and partial melting of billets prior to injection.

A more recent process has been developed at MIT (the SSR® process) [35], which is claimed to have added advantages. A spinning cold rod is introduced into the molten alloy just above the liquidus to generate very fine spherical particles (Fig. 4.1).

As the liquid cools rapidly to a temperature a few degrees below the liquidus, nucleation occurs on the cold rod with solid growing into the undercooled melt. Given the rapid heat removal and therefore rapid growth of solid into the melt, it is likely that the solid grows as a fine dendritic structure. In this thermal and flowing environment, the dendrite arms will rapidly coarsen and melt off, providing large quantities of fine spheroidal solid particles that are detected in quenched samples (Figs. 4.2 and 4.3).

David H. Kirkwood et al., *Semi-solid Processing of Alloys*,
Springer Series in Materials Science,
DOI 10.1007/978-3-642-00706-4_4, © Springer-Verlag Berlin Heidelberg 2010

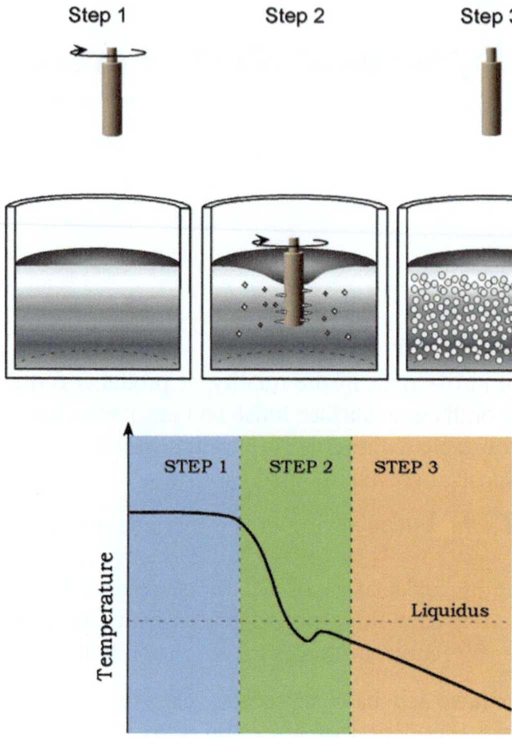

Fig. 4.1 Schematic of the SSR process

Fig. 4.2 Schematic of dendrite multiplication

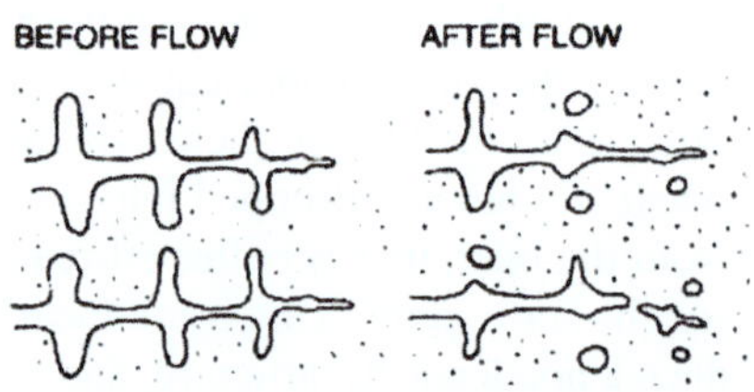

Fig. 4.3 Micrograph of Al–Cu alloy rheocast with SSR and then immediately quenched. Spheroidal primary grains are evident that formed during the initial period of solidification with subsequent dendritic microstructure that formed during the quench

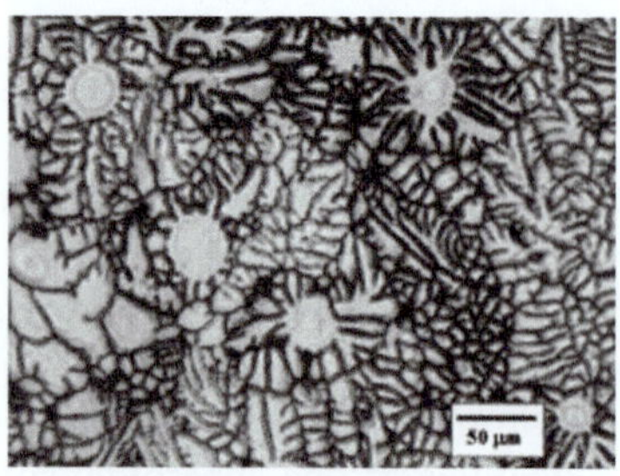

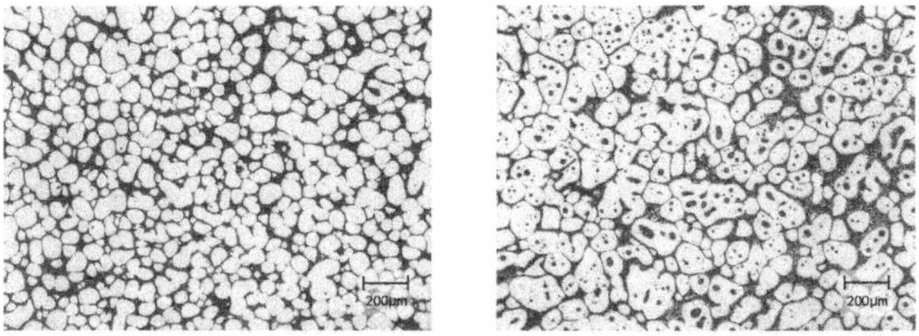

(a) (b)

Fig. 4.4 Microstructures of reheated and quenched Al–Si alloy produced by **(a)** SSR and **(b)** electromagnetically stirred MHD continuous cast billet [35]

A few degrees below the liquidus, where $f \cong 0.1$, the spinning rod is removed and the melt allowed to solidify more slowly in quiescent conditions. The spherical nuclei continue to grow as spheres to produce a fine grain structure of particle size of around 70 μm. This structure may be remelted, as in normal thixoforging practice, to produce a semisolid billet containing $f_s = 0.5$ for die injection, and given the fine grain size and lack of entrapped liquid within the solid particles, this slurry will flow easily to provide a casting of good mechanical properties (Fig. 4.4).

Alternatively, the solidification process may be halted at a higher temperature corresponding to $f_s = 0.3$; at this low fraction solid, the slurry flows almost like the liquid melt and may be poured into the small shot hole of a conventional die casting machine. This approach combines the advantages of thixoforming producing a fine structure and avoiding air entrapment that allows heat treatment, with the convenience of conventional die casting and also allowing scrap recycling in-house.

References for Part I Evolution and Design of Microstructure in Semisolid Alloys

1. M.C. Flemings, Metall. Trans. A **22A**, 957 (1991)
2. M.C. Flemings, *Solidification Processing* (McGraw-Hill Book, New York, NY, 1974)
3. W. Kurz, D.J. Fisher, *Fundamentals of Solidification*, 3rd edn. (Trans Tech, Switzerland, 1989)
4. J.H. Holloman, D. Turnbull, Prog. Met. Phys. **4**, 333 (1953)
5. A.L. Greer, Philos. Trans. R. Soc. Lond. A **361**, 479 (2003)
6. I. Maxwell, A. Hellawell, Acta. Metall. **23**, 229 (1975)
7. S.C. Flood, J.C. Hunt, *Metals Handbook*, vol 15, 9th edn. (ASMI, Ohio, USA, 1992), pp. 130–136
8. B. Chalmers, J. Aust. Inst. Met. **8**, 223 (1962)
9. A. Vogel, B. Cantor, J. Cryst. Growth **37**, 309 (1977)
10. B. Cantor, A. Vogel, J. Cryst. Growth **41**, 109 (1977)
11. Qin, Fan, in *6th International Conference On Semi-Solid Processing of Alloys, Composites*, Turin, 2000, ed. by G. Chiametta, E. Rosso, pp. 819–824
12. Z. Fan, Int. Mater. Rev. **47**, 49 (2002)
13. W.R. Loué, M. Suéry, Mater. Sci. Eng. A, **A203**, 1 (1995)
14. J. Gurland, Trans. Met. Soc. AIME **212**, 452 (1958)
15. J. Gurland, Trans. Met. Soc. AIME **236**, 624 (1966)
16. Lee, J. Gurland, Mater. Sci. Eng. **33**, 125 (1978)
17. E.A. Brandes, G.B.Brook, *Smithells Light Metals Handbook* (Butterworth-Heinemann, Oxford, 1988)
18. G-C. Gullo, K. Steinhoff, P.J. Uggowitzer, in *Proceedings of 6th International Conference on Semi-Solid Processing of Alloys and Composites*, Turin, 2000, ed. by Chiametta, Rosso, pp. 367–372
19. M.C. Flemings, Mater. Trans. (Jpn. Inst. Met.) **46**(5), 895 (2005)
20. D.H. Kirkwood, Mater. Sci. Eng. **73**, L1 (1985)
21. I.M. Lifshitz, V.V. Slyozov, I. Phys. Chem. Solids **19**, 35 (1961)
22. C. Wagner, Z. Electrochem. **65**, 581 (1961)
23. T.Z. Kattamis, J.C. Coughlin, M.C. Flemings, Metall. Trans. **239**, 1504 (1967)
24. P.W. Wilson, T.Z. Kattamis, Y.J. Shiohara, J. Mater. Sci. **23**, 2882(1988)
25. M. Chen, T.Z. Kattamis, Mater. Sci. Eng. A **247**, 239 (1998)
26. D.R. Poirier, S. Ganesan, M. Andrews, P. Ocansey, Mater. Sci. Eng. A **148**, 289 (1991)
27. K.H. Chien, T.Z. Kattamis, Z. Metallkunde **61**, 475 (1970)
28. K.P. Young, D.H. Kirkwood, Metall. Trans. **6A** 197 (1975)
29. E. Tzimas, A. Zavaliangos, Mater. Sci. Eng. A **289**, 217 (2000)
30. M.P. Kenney, J.A. Courtois, R.D. Evans, G.M. Farrior, C.P Kyonka, A.A. Koch, K.P. Young, *Metals Handbook*, vol 15, 9th edn. (ASMI, Ohio, USA, 1992), pp. 327–338
31. J. Alkemper, P.W. Voorheers, Acta. Mater. **49**, 897 (2001)
32. O. Ludwig, M. DiMichiel, P. Falus, L. Salvo, M. Suery, TMS
33. N. Limodin, L. Salvo, M. Suery, M. DiMichiel, Acta Mater. **55**, 3177(2007)
34. T. Courtney, Metall. Trans **8A**, 685 (1977)
35. M. Adachi et al., in *Trans. Japan Die Casting Congress*, Yokohama, 1998, p. 123
36. J.A. Yurko, R.A. Martinez, M.C. Flemings, in *Proceedings of 8th International Conference on Semi-Solid Processing of Alloys and Composites*, Cyprus, 2004, Paper 02–01

Part II
Rheology and Modeling

M. Suery and H. Atkinson

Chapter 5
Introduction and Definitions for Rheology and Modeling

The semisolid state in a metallic alloy can be obtained either during solidification from the liquid state, or during partial remelting from the solid state.

Metal forming during solidification (rheocasting) has been studied in the early stages of the development of semisolid processing by Flemings and his coworkers, but it was not the main forming method until very recently with the development of the new rheocasting (NRC) process. However, solidification of an alloy with mechanical, passive, or electromagnetic stirring was extensively used to produce globular microstructures. Therefore, knowledge of the rheological properties of alloys subjected to stirring during solidification is of great importance in comparison with non-stirred alloys on the one hand and with partially remelted alloys on the other hand.

Metal forming was thus mainly carried out after partial remelting: thixoforming (thixocasting or thixoforging) is then concerned with solid fractions, which can vary in a quite large range, from 10 to 20% in thixomolding to much larger values in thixoforging. Study of the rheological behavior of alloys during such a treatment is therefore also important for better forming conditions and improved properties of the formed components.

The alloys studied from a rheological point of view belong essentially to three categories: the model alloys like Sn–Pb with a low melting temperature for which the experiments are relatively easy, the aluminum and magnesium alloys, which are the main thixoformed alloys, and the alloys with a high melting temperature for which only feasibility tests have been carried out.

For the characterization of the rheological behavior of these semisolid alloys, two approaches are usually adopted.

The first approach is concerned with the semi-liquid state: the alloy contains low fractions of solid so that the behavior is described in terms of an apparent viscosity. The alloy is considered as a homogeneous medium, which is characterized during shear-dominated experiments. Fluidity measurements are also performed owing to the usefulness of this parameter for characterizing the flow behavior of the alloy during filling of a mould.

The second approach applies to the behavior at large solid fractions. The viscosity is no longer the pertinent parameter owing to the very important agglomeration phenomena, which are taking place between the solid particles. The alloy therefore

David H. Kirkwood et al., *Semi-solid Processing of Alloys*,
Springer Series in Materials Science,
DOI 10.1007/978-3-642-00706-4_5,

behaves more like a porous solid medium saturated with liquid. In this case, the behavior must be described by considering the two phases and their interactions.

Models are interesting if they can be easily integrated in computer simulation codes for optimizing the metal forming process in industrial production. Numerical simulation can indeed avoid long and expensive trial and error experiments by answering, well before the production stage, questions concerning the shape of the mould and the forming conditions. Several codes have already been developed for this purpose and others are still under development. These codes do not take into account all the complex and coupled physical phenomena that are occurring during a thixoforming or rheocasting operation but, even with simplifying assumptions, they can be very useful during the development stage of a new product.

Based on the previous considerations, this part will be divided into two main parts. The experimental determination of the rheological behavior of semisolid alloys will be first presented. The modeling of this rheological behavior will be then considered.

Chapter 6
Experimental Determination of Rheological Behavior

This chapter will be divided into two main parts. The behavior of alloys during partial solidification will be considered first but only alloys with a globular morphology will be examined. In this case, the alloy is considered as a homogeneous medium and therefore, the viscosity only (and sometimes the fluidity) has been determined. Alloys during partial remelting will thereafter be considered. In this case, the solid fraction can vary in large proportions so that the two approaches mentioned above will be detailed. In each part, the experimental methods will be examined first and then, the experimental results will be presented. It is necessary to mention here that only the most representative results will be presented and not all of the results in the literature.

In addition to these two main parts, a comparison between partially solidified and partially remelted alloys will be performed. This comparison is important in view of the latest industrial developments of the rheocasting processes. Finally, the question of yield stress will be addressed.

6.1 Partially Solidified Alloys

6.1.1 Experimental Methods

Rotation and Translation Viscometer

The Couette viscometer test has been used since the beginning of research concerning semisolid alloys for their rheological characterization [1, 2]. The advantage of this test is the possibility to generate the globular microstructure suitable for the forming operation directly in the apparatus. During the test, the alloy is sheared between two concentric cylinders, the inner cylinder being fixed and the outer cylinder rotating at constant angular velocity. A fixed outer cylinder and a rotating inner cylinder define a Searle viscometer. The torque, which is required to shear the alloy is then continuously recorded. From the angular velocity and the torque, it is possible to deduce the shear rate and the shear stress after some simple

David H. Kirkwood et al., *Semi-solid Processing of Alloys*, Springer Series in Materials Science,
DOI 10.1007/978-3-642-00706-4_6,

assumptions. The test is therefore very simple, but problems occur as soon as very clean experiments are required to produce quantitative data. The problems are concerned with:

- The nature of the materials used to build the apparatus owing to the aggressive behavior of particular alloys. This problem is particularly important in the case of aluminum alloys which dissolve steels. This requires either protective coatings on steels or the use of other materials like graphite.
- The control and homogeneity of the temperature: This aspect is particularly important in the case of semisolid alloys for which the viscosity is strongly dependent on solid fraction and therefore on temperature. Problems associated with temperature can have three origins: temperature control, temperature gradients in the device, and adiabatic heating of the alloy during shearing. The control of the temperature is usually carried out through one or several thermocouples, which are positioned as close as possible to the alloy, which can lead to a small difference with the temperature of the alloy. In a similar way, longitudinal temperature gradients can be present, which directly depend on the heating system. Viscous heating can be generated in the case of high shear rates. However, for aluminum alloys sheared at angular velocity of 500 rpm, it has been demonstrated that this heating does not exceed a few thousandths of a degree which is completely negligible [3].
- Turbulence effects: The equations which allow the determination of the viscosity in the viscometer assume laminar flow for the semisolid alloy. This assumption, which may be wrong in the case of liquid sheared at high velocities, is generally valid in the case of semisolid alloys for which the viscosity increases rapidly with increasing solid fraction.
- End effects: The viscometer has a finite length so that the torque must in principle be corrected to account for end effects. Indeed, a small part of the alloy is present under the inner cylinder and it is sheared under a variable velocity lower than that applied in between the cylinders. For alloys, which exhibit a strong shear thinning behavior, this shearing can induce a relatively important torque, all the more when the length of the viscometer is small.
- Wall slip and groove effects: The shear that is imposed on the alloy located between the two cylinders is assumed to vary linearly from the wall of the fixed cylinder to that of the rotating one. The alloy in contact with the rotating wall is therefore moving with the same velocity. However, slip can occur at this wall, which can lead to erroneous measurements. To avoid slip, longitudinal grooves are often machined along the two cylinders, which can cause nondesirable effects.
- Effects of variation of solid fraction: Good measurements assume that the material, which is sheared between the cylinders is homogeneous. This assumption is certainly valid in the case of a liquid, but it does not hold when particles are present as is the case for an alloy during solidification. It is, however, quite reasonable for particles that are small compared with the dimension of the gap between the cylinders. It must be noted that gravity or centrifugal acceleration can lead to heterogeneous distribution of the particles.

- Inertia effects: These effects can happen during rapid changes of shear rate and lead to some delay in the torque measurement. Extrapolation of curves is therefore necessary to obtain the material response at constant structure and this can lead to some error.

Beside the Couette viscometer, other rotation viscometers can be used: the cone–plane viscometer, the plane–plane viscometer or more simply viscometers which are constituted of a part rotating in the solidifying alloy. In this latter case, calibration with fluid of known viscosity is mandatory owing to the complicated geometry of the system.

In addition to rotation shear experiments, shear can be produced by translation of one part of the specimen with respect to the other. This test has been used for the first experiments dealing with the rheological behavior of semisolid alloys in order to establish relations between the strength of the alloy and its tendency to hot crack formation [4,5]. It consists of translating one part of the sample, the other part being fixed. Shear then occurs in a very narrow region so that it is very difficult to know precisely the imposed strain. This test has been taken up again more recently on Al–Si alloys [6] and even more recently on Al–Si–Cu alloys [7]. Another possible configuration of the translation shear test is the use of two concentric cylinders with circumferential grooves. As in the Couette viscometer, the alloy is located in between the cylinders but a translation is imposed to one cylinder with respect to the other. For small displacement, pure shear is imposed on the alloy over a known distance so that strain is also known, provided that no slip occurs at the wall. This apparatus has been used for the study of aluminum alloys during solidification at particularly large solid volume fractions [8]. In the following, the results obtained with these viscometers will not be reported since they are mainly concerned with dendritic structures during solidification and transitions in the behavior when the solid fraction increases.

Shearing of a semisolid alloy can be produced by parallel plate compression. Although this type of test is better suited for partially remelted billets, it has been employed also with billets transferred from the cooling system to the parallel plate test machine [9, 10]. In these experiments, a dead weight is applied on the specimen and the displacement is recorded continuously, the billet temperature being kept constant by using a furnace installed on the press.

Fluidity Tests

Fluidity tests have been used to characterize the rheological behavior of semisolid alloys particularly at relatively low solid fractions. Fluidity does not have a mathematical definition, in contrast with viscosity. It is used by foundry-men and consists of the measure of the length that an alloy travels in a tube of a given geometry before it solidifies. This is therefore a comparative method, which can be carried out for the liquid as well as for the semisolid alloy. Two systems are usually used but it is possible to imagine other ones. The first (Fig. 6.1a) is constituted of a horizontal spiral and a vertical sprue. The liquid or semisolid alloy is poured inside the sprue,

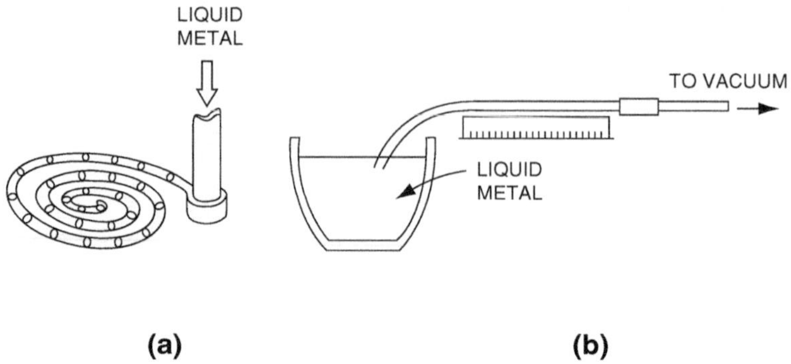

Fig. 6.1 Devices used to measure the fluidity. (**a**) spiral; (**b**) vacuum suction

which travels a given distance in the spiral. In the second apparatus (Fig. 6.1b), the alloy is held in a crucible and a tube connected with a vacuum system is submerged in the alloy. In this case too, the distance that the alloy has traveled in the tube is recorded as the fluidity parameter.

6.1.2 Results Concerning Viscosity

The first results concerning viscosity of alloys during solidification have been obtained at MIT with a Sn–Pb alloy [1]. These authors measured the apparent viscosity of the alloy during solidification with mechanical stirring. The results showed that the stirred semisolid alloy at solid fractions greater than 0.2 behaves like a non-Newtonian fluid with a much smaller viscosity than the same unstirred alloy. This first observation is at the origin of the very large number of studies dealing with the rheology of stirred alloys. The very important work of Joly and Mehrabian [2] on the Sn–Pb system too is among these studies. Today, although numerous studies have been carried out on other systems and have tried to understand more complex aspects of the rheological behavior, these works are still the basis of our knowledge of this behavior. Three main aspects of the variation of the apparent viscosity of these alloys have been studied:

- The variation of the viscosity at constant shear rate during continuous cooling
- The variation of viscosity at constant temperature as a function of the shear rate
- The thixotropic behavior of the alloy, i.e., the time dependence of the rheology

The determination of the apparent viscosity of an alloy during continuous cooling can be viewed as a very simple and fast characterization method of the rheology. The test is carried out at constant shear rate so that the increase of viscosity is due to the increase of the solid fraction. The effects of the shear rate and of the cooling rate can also be studied during these experiments. The classical curves giving the variation of the viscosity as a function of the solid fraction show that the increase of the shear rate increases the critical solid fraction at which the viscosity increases sharply (Fig. 6.2). Similar curves have been obtained on other semisolid alloys like industrial Al–Si alloys and on alloys containing SiC particles.

In order to express the influence of the solid fraction f_s on the viscosity η of the alloy, laws of the type $\eta = A \exp(Bf_s)$ have been used. Table 6.1 gives the values of B deduced from viscometer experiments on various alloys. In this table,

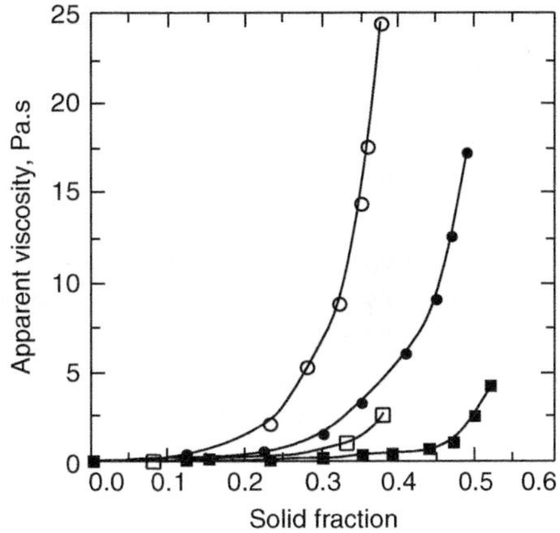

Fig. 6.2 Apparent viscosity as a function of the solid fraction for an A356 alloy sheared at (*filled square*) 27 s^{-1}, (*open square*) 54 s^{-1}, (*filled circle*) 108 s^{-1}, and (*open circle*) 216 s^{-1} during continuous cooling in the solidification interval at 1.2°C min^{-1} (from [11])

Table 6.1 Values of B in the expression: $\eta = A \exp(Bf_s)$ for various alloys

Alloy	Conditions of the tests	Value of B	References
Sn-15%Pb	Continuous cooling at 0.33°C min^{-1} Shear rate of 750 s^{-1}	7.4	[2]
Sn-15%Pb	Continuous cooling at 25°C min^{-1} Shear rate of 750 s^{-1}	17.6	[2]
A356	Continuous cooling at 1.2°C min^{-1} Shear rate of 27 to 216 s^{-1}	15	[11]
A356	Isothermal holding Shear rate of 27 to 216 s^{-1}	11.3	[11]
A356	Continuous cooling at 15°C min^{-1} Shear rate of 0.063 s^{-1}	10.7	[3]
A356	Continuous cooling at 15°C min^{-1} Shear rate of 105 s^{-1}	15.6	[3]
A356 + 18% SiCp	Continuous cooling at 15°C min^{-1} Shear rate of 105 s^{-1}	16.9	[3]
A356	Isothermal holding Shear rate of 1.6 to 105 s^{-1}	9.5	[3]
A356 + 18% SiCp	Isothermal holding Shear rate of 26 to 210 s^{-1}	11.7	[3]

values deduced from isothermal experiments are also included together with results on composites.

The results show a very strong dependence of the viscosity on the solid fraction with B values ranging between 10 and 18. Although it is difficult to derive clear tendencies, it seems that the B value is smaller for low cooling rates and for isothermal holding, which can be considered in fact as a cooling at zero rate. The interpretation that can be put forward for these large B values is the presence of agglomerates of solid globules, which entrap part of the liquid. The liquid fraction participating in the flow can be therefore considerably reduced and this occurs all the more when the agglomerates are formed at high cooling rates.

A similar type of law was derived [10] for the influence of the aspect ratio AR of the globules on the viscosity of A356 Al–Si alloys deduced from parallel plate compression viscometry at shear rates lower than $0.01\,s^{-1}$. For a solid fraction of 0.36, they found that the viscosity increases exponentially with increasing aspect ratio according to: $\eta = K \exp(11.28\,\mathrm{AR})$ for values of AR ranging from 1.4 to 1.8. This law corresponds to a B coefficient ranging from about 44 for AR = 1.4 to about 56 for AR = 1.8. These values are much greater than the values given in Table 6.1, which can be possibly explained by the very low strain rates applied during these compression experiments for which agglomeration of the solid globules is important.

Shear experiments during isothermal holding allow a more precise rheological characterization. They consist of a partial solidification of the alloy at a given shear rate, until a steady shear stress is reached. Then, another shear rate is applied and the response in terms of shear stress is recorded. The main characteristic feature of these experiments is that the measure is performed after sufficient time to reach steady state. This time can be quite long as shown in Fig. 6.3 obtained on an Al–Si alloy A356 [11].

Similar experiments have been carried out by Kattamis and Piccone [12], on an Al–4.5%Cu–1.5%Mg, but at various solid fractions. Figure 6.4 shows the variation

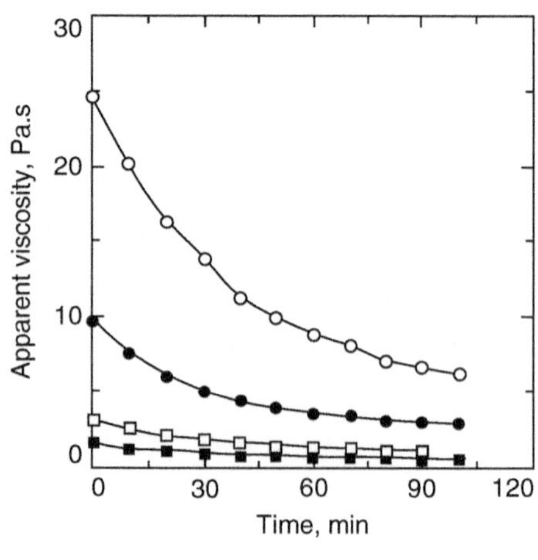

Fig. 6.3 Apparent viscosity as a function of shearing time at 590°C of a A356 alloy ($f_s = 0.35$) partially solidified with a shear rate of (*filled square*) $27\,s^{-1}$, (*open square*) $54\,s^{-1}$, (*filled circle*) $108\,s^{-1}$, and (*open circle*) $216\,s^{-1}$ (from [11])

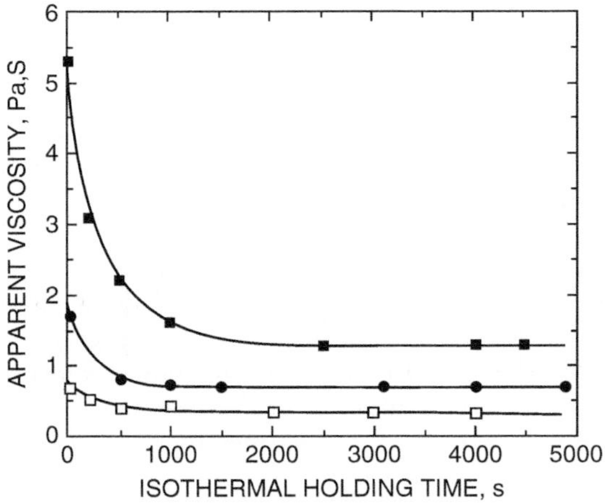

Fig. 6.4 Apparent viscosity as a function of shearing time at a shear rate of $200\,s^{-1}$ and at various temperatures corresponding to solid fraction $f_s = 0.3$ (*open square*), 0.4 (*filled circle*), and 0.5 (*filled square*). Al-4.5%Cu-1.5%Mg (from [12])

of the apparent viscosity as a function of shearing time at a constant shear rate of $200\,s^{-1}$ and at various temperatures corresponding to various solid fractions. Steady state viscosity is reached after 1,000–2,000 s depending on the solid fraction.

Steady state viscosity was not observed in the case of the AZ91D magnesium alloy [13] sheared at $120\,s^{-1}$ at different temperatures corresponding to different solid fractions from 0.429 to 0.5. As the stirring time increases, the apparent viscosity initially decreases until it reaches a minimum value after which it starts to increase. This increase was attributed to the growth of the dendritic particles into spherical shapes.

In steady state, the rheological behavior has been found to be strongly shear thinning, the apparent viscosity decreasing with increasing shear rate (Fig. 6.5). Such a behavior has been confirmed on other semisolid systems and for higher shear rates. However, a tendency to observe saturation in the decrease of the viscosity with the highest shear rates has been reported [14].

The explanation for the shear thinning behavior of semisolid alloys is the decrease of the agglomerate size of the solid particles when the shear rate increases by rupture of the solid bridges. This decrease of the size leads to the release of the entrapped liquid, thus reducing the viscosity. Such an explanation has been confirmed at least partially by the work of Ito et al. [15] carried out on Al–Si alloy sheared at various rates. These authors confirmed that the concept of agglomerated particles is entirely justified through metallographic observations. In addition, they confirmed that high shear rates lead to a smaller volume fraction of entrapped liquid. More recently, observations carried out by X-ray tomography have confirmed that semisolid structures are highly agglomerated as soon as the solid fraction exceeds 0.3 or 0.4 [16]. Indeed, solid bridges easily form between slightly misoriented particles producing

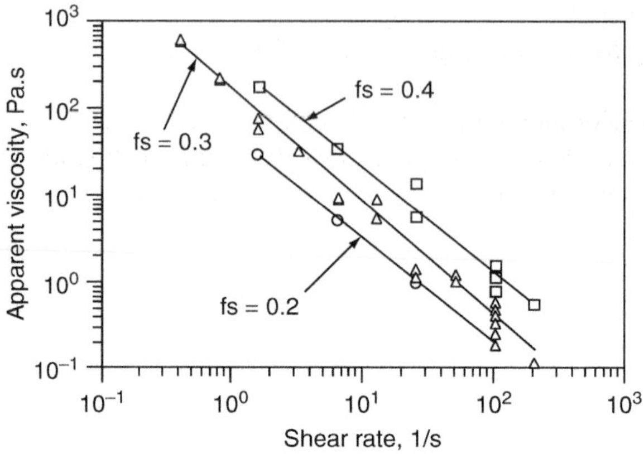

Fig. 6.5 Apparent viscosity of an A356 alloy as a function of the shear rate for various solid fractions (from [3])

a low angle grain boundary, which is not wetted by the liquid. The increase of the shear rate can therefore easily break the bridges so that the steady viscosity is the result of a dynamic equilibrium between the formation of bridges by contact of particles and diffusion, and their rupture induced by the shearing process.

The analysis of the slope of the curves reported in Fig. 6.5 shows that it is smaller than -1 (close to -1.3), which means that the shear stress decreases with increasing shear rate. Similar results were observed by other authors on Sn–Pb alloys [17, 18]. In their experiments, the steady state shear stress values were also measured by step changes in shear rate with sufficient time at each rate to reach a steady value of the torque. For McLelland et al., this anomalous behavior is interpreted in terms of rapid structural breakdown of particle agglomerates on increasing the shear rate from low initial values. For Koke and Modigell, this behavior can be explained at least partly by the change of the size of the globules. Indeed, when step changes in shear rate are carried out, the particle size is larger at the end of the experiment compared to the beginning and this explains at least partly the decrease in shear stress. These experiments therefore suggest that the rheological behavior at steady state depends on the way the material is tested.

Very recently, Kirkwood and Ward [19] demonstrated that the slope of the curve, such as that shown in Fig. 6.5 should be equal to $-4/3$. This value is proposed in conformity with the principle of shear reversal and it is independent of alloy system and fraction solid.

The viscosity at steady state has been related to the effective solid fraction [15], the effective solid fraction taking into account the liquid fraction entrapped in the agglomerates. Figure 6.6 shows that this viscosity depends only on this effective solid fraction without any influence of the shear rate. This result is not contrary to the idea of shear thinning since a similar value of the steady state viscosity can be obtained at two different shear rates corresponding to different volume fractions of entrapped liquid. The analysis of the straight line obtained for effective solid

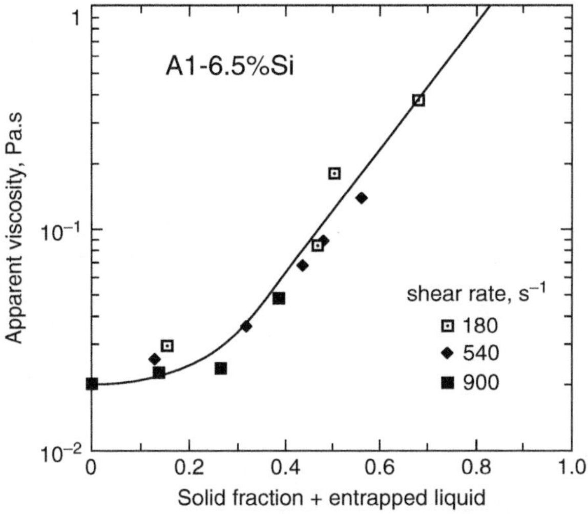

Fig. 6.6 Variation of the apparent viscosity at steady state as a function of the volume fraction of solid increased by the liquid volume fraction entrapped in the agglomerates for various shear rates (from [15])

fractions between 0.2 and 0.7 leads to a B value of 6.3, which is much smaller than that, about 10, obtained when the real solid fraction is considered (Table 6.1).

Another important aspect of the work of Joly and Mehrabian [2] is the study of the thixotropic behavior of semisolid Sn–Pb alloys. Thixotropy is defined as the time dependency of the viscosity. It reveals itself by the occurrence of hysteresis loops in the shear stress–shear rate diagram, when the shear rate is decreased until zero and increased again to its initial value after a given resting time. Figure 6.7 from the work of Joly and Mehrabian [2] shows such curves after various resting times. It is interesting to note that all the curves come back to the same initial point showing thus the reversible character of the microstructure of semisolid alloys. This result is obtained because the duration of the experiments is very short so that coarsening of the globules is not significant. In addition, coarsening of the globules is not a very important factor influencing the shear stress compared with the degree of agglomeration, provided that the size of the globules remains much smaller than the size of the specimen which is tested.

Other authors have used a similar procedure in the case of aluminum alloys containing SiC particles [20]. However, this procedure is not sufficient to fully characterize the agglomeration and deagglomeration kinetics, which are at the origin of the thixotropic behavior of semisolid slurries.

The transient evolution of stress observed after a step change of strain rate is a means to characterize these kinetics, provided that the tests are performed in very well controlled conditions (no slip at the walls during the shear rate change, rapid acquisition of the data, no inertial effects). Agglomeration processes are then dominant during step changes with decreasing shear rate values, whereas

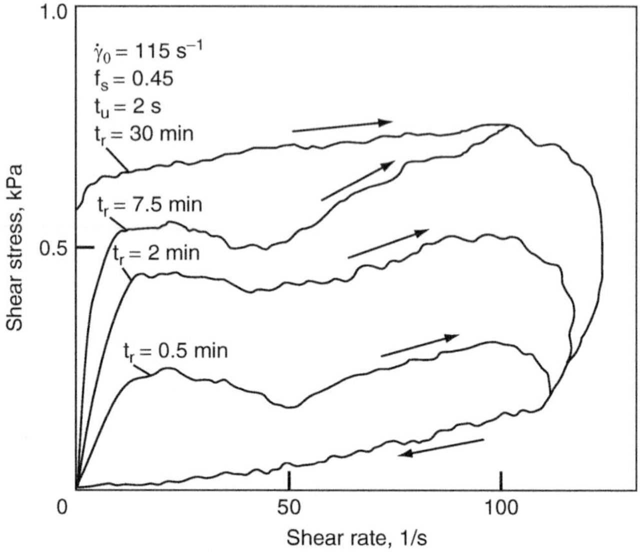

Fig. 6.7 Effect of the resting time t_r on the hysteresis loops of a Sn-15%Pb alloy; t_u is the time required to increase the shear rate up to its maximum value (from [2])

deagglomeration becomes dominating with increasing shear rate. Tests conducted in this way have shown that the alloy can exhibit a shear thickening behavior, viscosity increasing with increasing shear rate [21]. This shear thickening behavior was confirmed by Koke and Modigell [18], who studied the viscosity of a Sn-15.8%Pb alloy by using a Searle-type rheometer. By excluding inertial effects from the data evaluation, they obtained an isostructural flow curve with a shear stress depending on the shear rate to the power 2.07. The viscosity therefore increases almost linearly with increasing shear rate. Gautham and Kapur [22] observed also a shear thickening behavior under isostructural conditions, but the power law exponent was smaller, equal to 1.2.

The effect of the rest time on the shear stress measured during shear rates jumps has been determined by various authors [22, 23]. During the experiments, the shear rate was increased from 0 to a value typical of that experienced by the alloy when it enters the die cavity during an industrial thixoforming process. Increasing rest times lead to an increase in agglomeration and particle sizes thus increasing the shear stress recorded when the material is subjected to a shear rate jump from rest (Fig. 6.8). In addition, it was observed that the transient behavior of the slurry after a rapid change in shear rate occurs within 1 s of the shear-rate jump [23].

The increase of the resting time does not always lead to an increase of viscosity at a given shear rate. In the case of the AZ91D alloy, an increase of the viscosity was observed at high solid fraction (0.4), whereas at lower solid fraction, viscosity decreases slightly with increasing resting time [13]. The authors explained this result by different coarsening mechanisms. At high solid fraction, particle

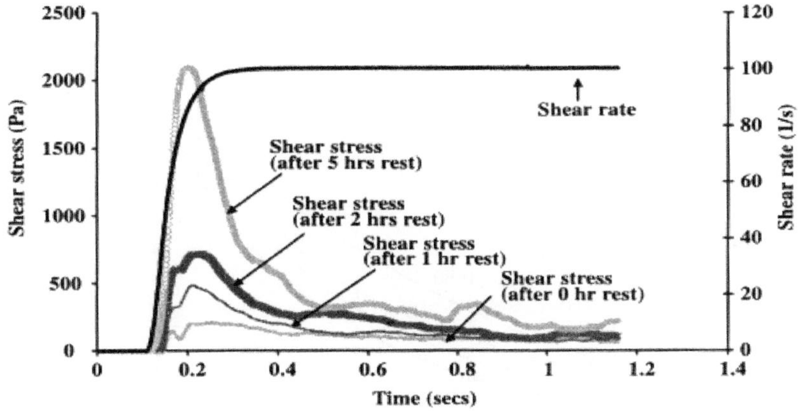

Fig. 6.8 Shear rate jumps from 0 to 100 s^{-1} after different rest times for Sn-15%Pb alloy at fraction solid 0.36 [23]

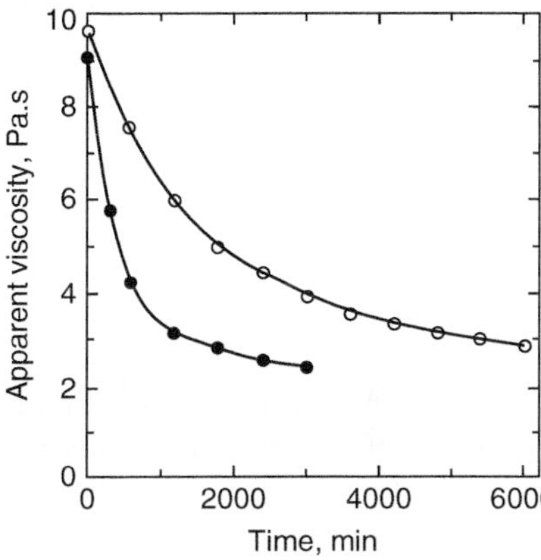

Fig. 6.9 Apparent viscosity as a function of the isothermal shearing time of an A356 alloy partially solidified ($f_s = 0.35$), (*open circle*) without and (*filled circle*) with 15% SiC particles added at time $t = 0$. The shear rate is 108 s^{-1} (from [11])

coalescence occurs preferentially to form larger ones in which the liquid entrapped in the individual particles remains entrapped. When the solid fraction is lower, the solid particles coarsen by Ostwald ripening and release the entrapped liquid into the bulk liquid. The effective liquid fraction is thus larger, which explains the viscosity decrease.

The results presented previously are concerned with semisolid alloys. Studies have been also carried out on alloys containing ceramic particles leading to similar results. However, it has been demonstrated that for particle fractions that are not too high, the viscosity of the alloy containing the ceramic particles can be lower than that of the alloy alone at the same temperature despite the presence of the particles (Fig. 6.9) [3, 11, 24]. This lower viscosity at higher solid fraction can be explained by the lower agglomeration level of the solid globules owing to the presence of the ceramic particles.

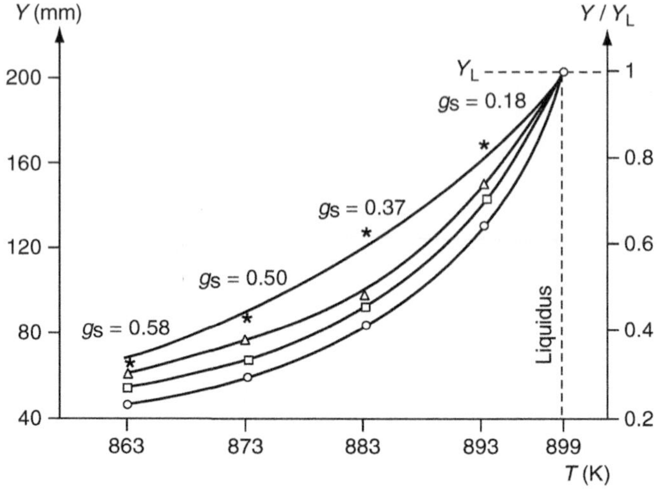

Fig. 6.10 Fluidity Y of an Al-10%Cu alloy in the semisolid state characterized by a solid fraction g_s. Y_L represents the fluidity of the liquid. The various curves are concerned with various stirring rates in revolutions/min. *open circle*: 340; *open square*: 482; *open triangle*: 695; *star sqaure*: 992 (from [25])

6.1.3 Results on Fluidity

As previously mentioned, the fluidity of an alloy is a technological parameter used by the foundry-man, who is pouring the metal in a mould to produce a part. Fluidity tests have been carried out with a partially solidified alloy subjected to stirring in comparison with the same alloy in the liquid state. The tests have shown that the alloy containing 50% solid still exhibits a relatively high fluidity and it is higher when the stirring rate before the test is increased [25] (Fig. 6.10). This result is in agreement with the shear thinning behavior of the alloy, the fluidity increasing (viscosity decreasing) with decreasing shear rate.

6.2 Partially Remelted Alloys

6.2.1 Experimental Methods

Shear Viscometer

During the shape forming of a billet by thixocasting, it experiences a very intense shear at the gate so that the capillary viscometer can be very interesting to obtain data about the rheological behavior of the billet in conditions close to forming. The capillary viscometer is similar to an extrusion apparatus in which the die is very

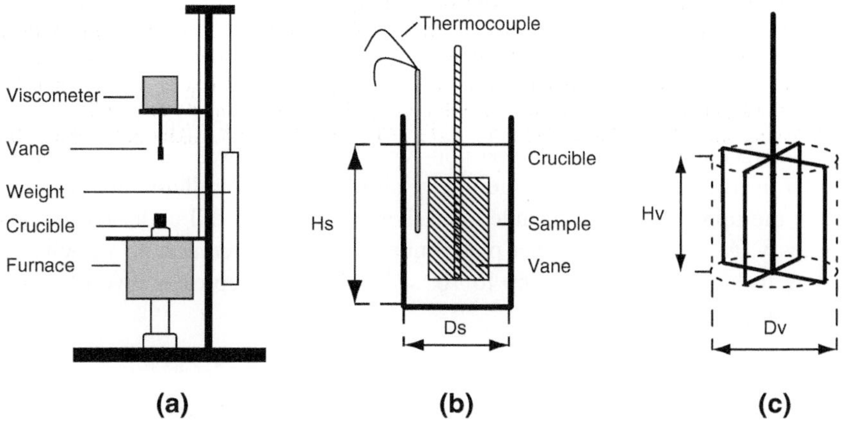

Fig. 6.11 Schematic drawing of the vane viscometer; (**a**) complete viscometer; (**b**) crucible with thermocouple and vane; (**c**) four bladed vane

wide. The pressure required to make the alloy flow in the tube at a given velocity is recorded and pressure sensors located along the tube allow the measurement of the pressure variations along the tube linked to the fluid flow. Such a device has been used by Paradies et al. [26] and Bernhard et al. [27].

The vane viscometer has been developed at NTNU Trondheim, Norway [28]. In a crucible placed in a furnace, the studied alloy is subjected to shear through the rotation of a four bladed vane (Fig. 6.11). This device can be used during solidification starting with the liquid alloy but it has been used mainly during partial remelting starting with the fully solid alloy. In this case, the specimen must be machined in order to allow the blades to be inserted in the alloy. The test consists of either imposing a shear at a given velocity, or imposing a stress during heating and recording the temperature (i.e., the liquid fraction) at which the material starts to deform.

Parallel Plate Compression

Parallel plate compression tests have been widely used to characterize the rheological behavior of semisolid alloys owing to the simplicity of the test. However, because friction cannot completely be avoided between the sample and the plates, the stress state is not purely uniaxial so that liquid segregation phenomena occur.

Two variants of the test have been used: the diameter of the specimen is smaller than that of the plates so that the current cross section of the specimen increases during the test or the specimen diameter is equal to that of the plates and in this case, the cross section remains constant which simplifies the analysis.

The problem of the flow of a fluid between two parallel plates has been solved many years ago. The solution is presented in [29] for the example of a Newtonian fluid, assuming sticky contact between the fluid and the plates. The Stefan equation is then obtained:

$$F = \frac{3\pi\eta R^4}{2h^3}\frac{\mathrm{d}h}{\mathrm{d}t}, \tag{6.1}$$

where F is the compression force, h is the instantaneous height of the specimen, R its radius, η is the viscosity of the fluid and $\frac{\mathrm{d}h}{\mathrm{d}t}$ is the compression rate.

More recently, a novel compression-type viscometer was developed [30–32]. The Drop forge viscometer has geometry similar to the parallel-plate viscometer, except the upper plate is suspended and then permitted to fall under the influence of gravity. As a result, the upper plate impacts the specimen at high velocity, thus leading to high shear rates of up to $10^4\,\mathrm{s}^{-1}$, which are representative of forming operations.

Direct Extrusion

The direct extrusion test consists of forcing a specimen of a given cross section to go through a die of smaller cross section under the action of a piston driven at a given velocity. This test is similar to the capillary viscometer and it reproduces quite well the conditions prevailing during thixocasting. The main drawback of the test is the friction between the piston and the container, which does not allow a precise determination of the rheology of the alloy. Nevertheless, experimental conditions leading to homogeneous deformation or liquid segregation can be defined by this test.

Back Extrusion

The previously mentioned drawback concerning friction is avoided if the test is performed in an inverse manner. The piston then plunges into the specimen, which flows between the container and the piston. This test is also close to industrial forming conditions and it allows the definition of the conditions for liquid segregation [33, 34]. Hence, the rheological behavior of the alloy in terms of apparent viscosity can be found, provided that some assumptions are made about the material and the boundary conditions [11].

Indentation

The indentation test used for characterizing semisolid alloys consists in the penetration of a tip of a given geometry into the specimen. It has been used at the beginning of the work on thixocasting to evaluate the softness of the material before starting the injection. It is therefore very interesting from this viewpoint, but it cannot easily lead to the rheology of the alloy in view of the very complex stress states that are generated under and in the vicinity of the tip. The analysis can be made by an inverse method using a computer code for the numerical simulation of the test [35, 36].

Drained Compression

In order to evaluate the compressibility of the solid, drained oedometric compression tests and drained triaxial compression tests have been carried out [37]. The drained oedometric compression consists of compressing the specimen in a container against a filter, which is very permeable for the liquid (Fig. 6.12). The strain state is well defined but this is not the case for the stress state, which can lead to some problems during the analysis of the test. The test allows also the measurement of the liquid volume fraction, which is really entrapped inside the solid. In addition, after complete filtration, a specimen is obtained, which has the composition of the solid phase in the semisolid alloy. The test can be performed without any major problem with industrial alloys, i.e., with alloys of relatively high solidus temperature.

The poor knowledge of the stress state in the previous test does not hold in the drained triaxial compression test, since the stress state is imposed and the volumetric strain of the sample is directly determined through the liquid quantity, which flows across the filter (Fig. 6.13). However, the specimen must be wrapped into a deforming envelope, which leads to problems in the case of industrial alloys. The

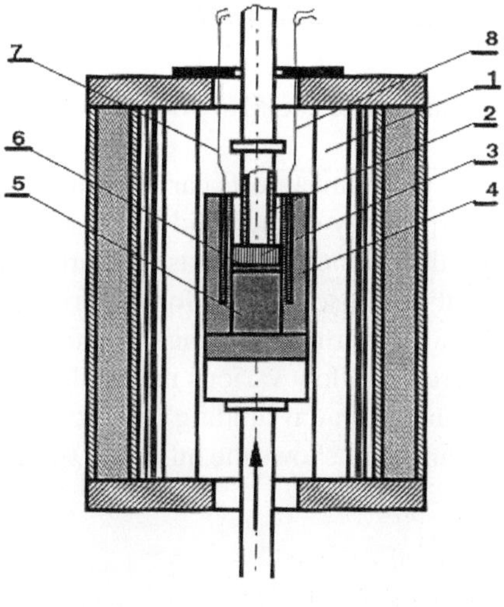

Fig. 6.12 Schematic drawing of the drained oedometric compression apparatus used for Al alloys. From [37]. 1: I.R. lamp furnace; 2: piston; 3: filter support; 4: container; 5: specimen; 6: fabric of SiC Nicalon fibers; 7, 8: thermocouples

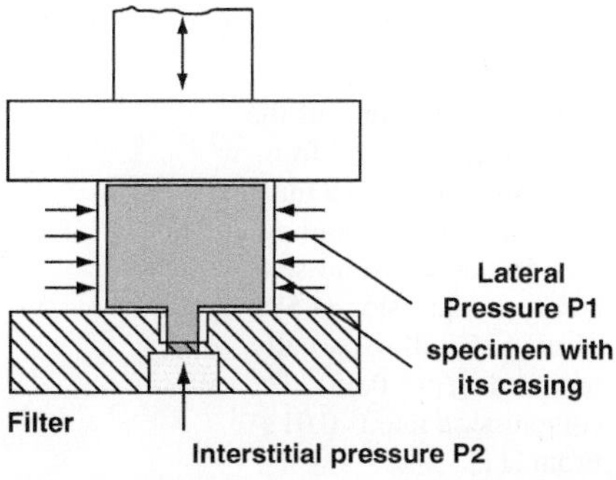

Fig. 6.13 Schematic drawing of the drained triaxial compression apparatus. From [37]

test has been applied successfully on low melting temperature alloys, such as Sn–Pb, the envelope being constituted of silicone rubber [37, 38]. It has also been used with the aluminum alloy A356. In this case, the envelope was made with pure Al and the pressure was applied by inert gas. The contribution of the envelope is then not negligible, which requires corrections of the applied axial stress [37].

Tension

The tensile test is normally not used in the case of semisolid alloys with a globular structure owing to the very low cohesion of the material when the solid fraction is typically that of billets suitable for thixocasting. However, in order to predict the liquid segregation phenomena, which can occur during a test, it is necessary to know the rheological behavior of the alloy for any stress states and higher solid fractions than those under normal conditions. Tensile tests have been carried out with magnesium alloys in particular [39].

6.2.2 Results in Terms of Apparent Viscosity

Influence of the Morphology of the Solid Phase

The mechanical tests carried out during partial remelting allow the influence of the morphology of the solid phase on the rheological behavior of the alloy to be easily determined. The tests are carried out, either after a predetermined holding time in the semisolid state for materials produced by various methods or subjected to various thermomechanical treatments, or after various holding times. These tests therefore allow various materials to be compared or they permit the determination of the optimum holding time in the semisolid state before the forming operation. Figure 6.14 shows the influence of the holding time on the shear stress deduced from

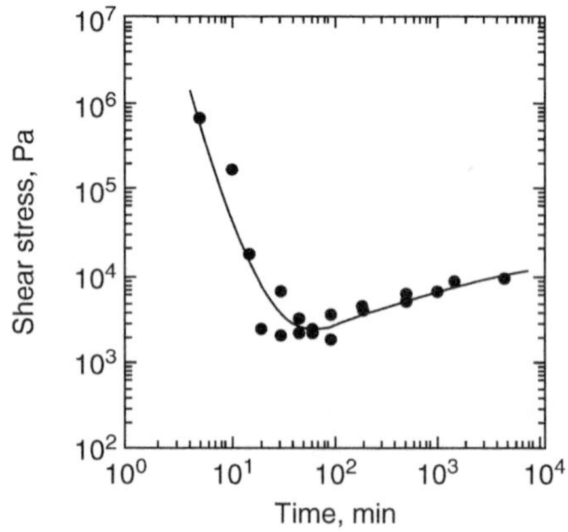

Fig. 6.14 Variation of the shear stress deduced from a compression test as a function of the isothermal holding time in the semisolid state before compression at 580°C of an A357 alloy partially remelted ($f_s = 0.45$). The compression rate is $0.01\ s^{-1}$. From [11]

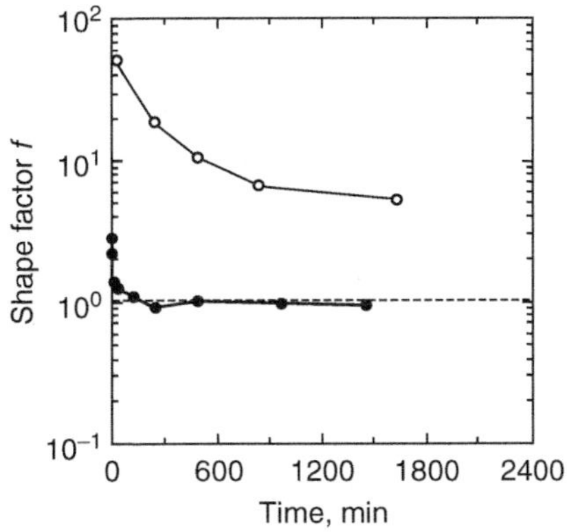

Fig. 6.15 Variation of the shape factor of the globule as a function of the isothermal holding time in the semisolid state at 580°C ($f_s = 0.45$) of an A357 alloy continuously cast (*open circle*) without and (*filled circle*) with electromagnetic stirring. The *dashed line* corresponds to perfect spheres. From [11]

a compression experiment carried out on an A357 aluminum alloy solidified initially with electromagnetic stirring. The shear stress (or viscosity) decreases sharply at short holding times to a minimum and then increases slightly when the holding time increases too much. The decrease of the stress is correlated with the globularization of the solid phase as shown by the concomitant decrease of the shape factor of the globules with increasing holding time (Fig. 6.15). This shape factor F is defined as:

$$F = \frac{1}{6\pi f_s} \frac{S_V^2}{N_A}, \tag{6.2}$$

where S_V is the globule–liquid interface surface area per unit volume and N_A represents the number of globules per unit area of section of the specimen. For long holding times, the increase of the viscosity is due to the agglomeration and growth of the globules, which become predominant over the globularization.

The entrapped liquid inside the globules plays a very important role in determining the viscosity of the alloy. This liquid does not participate in the rearrangement of the globules during deformation [15]. The alloy is thus equivalent to a material with a smaller liquid fraction, which therefore increases the viscosity. This increase is generally quite important owing to the very large sensitivity of the viscosity to the liquid fraction.

The degree of connectivity of the solid globules is also an important factor to take into account to describe the behavior of semisolid alloys. This connectivity can be quantified by the contiguity of the solid phase C_s or the contiguity volume $C_s f_s$ [40]. C_s is defined as $2N_L^{ss}/\left(2N_L^{ss} + N_L^{sl}\right)$ where $2N_L^{ss}$ is the number of interfaces between two solid globules per unit length and N_L^{sl} is the number of interfaces between the solid and the liquid per unit length (see the Microstructure Sect. 2.3, for a more complete discussion concerning contiguity). In the case of the electromagnetically stirred AA6082 tested by means of backward extrusion, these authors have shown that homogeneous deformation occurs when the contiguity volume is smaller than 0.3, whereas for larger values, the alloy loses its favorable flow characteristics. The influence of Ba addition was also studied and it was shown that Ba reduces the solid–liquid interface energy leading to an increased penetration of the liquid phase between globules.

Influence of Shear Rate

Shear thinning steady state behavior

As for the alloys tested during solidification that exhibit a shear thinning behavior, the partially remelted alloys show a viscosity, which decreases as the shear rate increases. Such an evolution can be observed over a very wide range of shear rates as shown in Fig. 6.16 obtained with aluminum alloys and a magnesium alloy by compression tests at low shear rates and back extrusion tests at high shear rates [39]. The slope of the curve is close to –1 for the aluminum alloys, which means that the shear stress is almost independent of the shear rate. The slope is a little higher for the magnesium alloy. These curves are characteristic of a steady behavior, which results for the stirred alloys from equilibrium between agglomeration and deagglomeration of the globules.

Transient behavior

In order to determine the behavior during transient situations, step changes of strain rates can be performed during simple compression experiments. Figure 6.17 shows

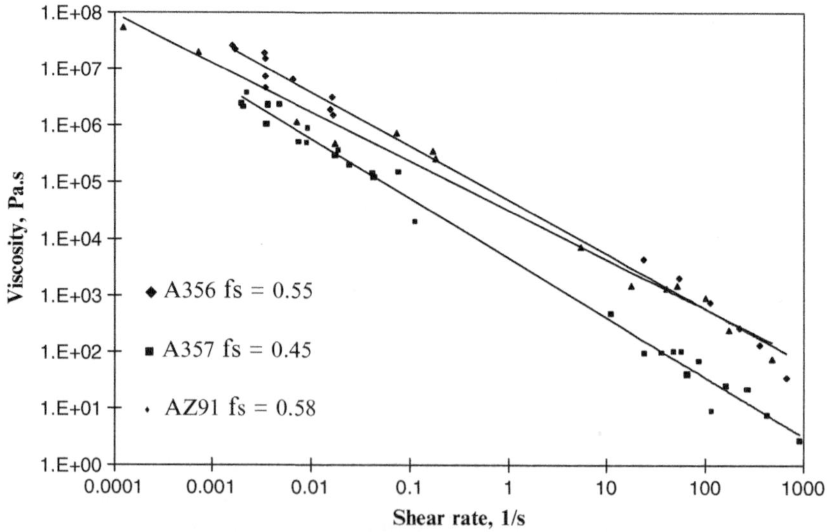

Fig. 6.16 Viscosity as a function of the shear rate for two aluminum alloys (A356, A357) and a magnesium alloy (AZ91) showing the shear thinning behavior of the alloys at solid fractions f_s close to 0.50. From [39]

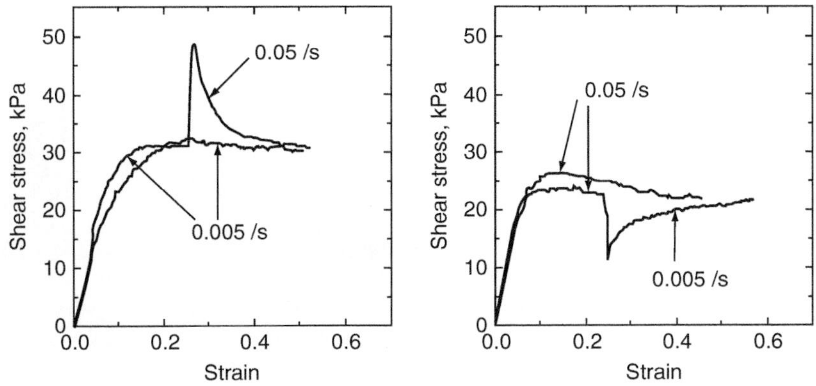

Fig. 6.17 Step changes of strain rate with increasing values (*left*) and decreasing values (*right*) for a Al-6%Si-0.6%Mg alloy deformed in compression at 580°C leading to a solid fraction of 0.55. From [11]

typical stress–strain curves resulting from step changes of strain rates with increasing and decreasing values. During an increase of strain rate, the stress increases rapidly and then decreases to a value close to that without strain rate change. Similarly, for a decrease of strain rate, the stress decreases sharply and then increases to the value close to that without step change. Such a behavior is in agreement with the previous result, which indicates that stress is almost independent of strain rate at steady state in this particular case. Conversely, the instantaneous change of the stress is characteristic of the behavior of the alloy at constant structure, i.e., at the degree of agglomeration of the globules, which is fixed by the strain rate applied before the strain rate change. The stress variation that follows immediately suppressed the step change is due to the deagglomeration of the globules for an increasing change and to the agglomeration for a decreasing change. As shown in Fig. 6.17, the strain for breakdown is smaller than that for buildup which is expected since breaking up of bonds between spheroidal solid particles in agglomerates is likely to be easier than the formation of bonds during shear rate drops.

The explanation in terms of microstructural changes is confirmed by tests with step changes of strain rate carried out on nonglobular structures on the one hand, and on aluminum matrix composites on the other hand. In the case of nonglobular structures, the stress increases during the strain rate change and shows a plateau characteristic of the new strain rate. The strain rate sensitivity of the stress is then characteristic of high temperature deformation of a solid.

A very similar behavior has been observed in the case of aluminum matrix composites containing SiC particles [42]. In these materials, the globular structure of the solid phase is obtained without any particular treatment owing to the presence of the SiC particles. Indeed, the particles are preferentially located in the liquid when the alloy is in the semisolid state, which allows a completely deagglomerated state of the structure (Fig. 6.18 [41]). It is to be noted, however, that this particular position of the particles is only possible for solid fractions, which are not too high

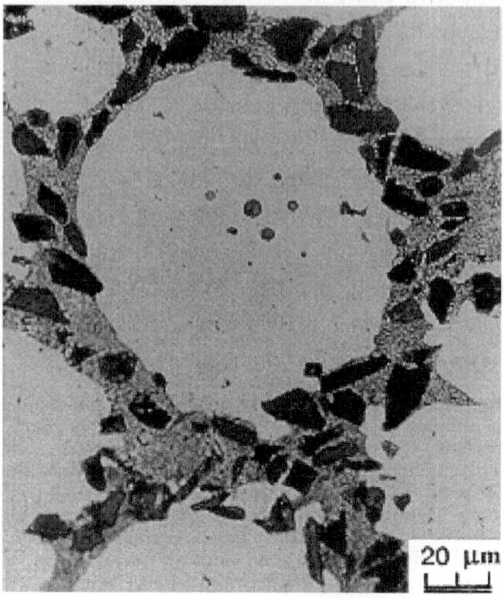

Fig. 6.18 Micrograph of a A356 + 20%SiC composite after partial remelting and quenching showing the position of the SiC particles in the eutectic mixture (liquid in the semisolid state). From [41]

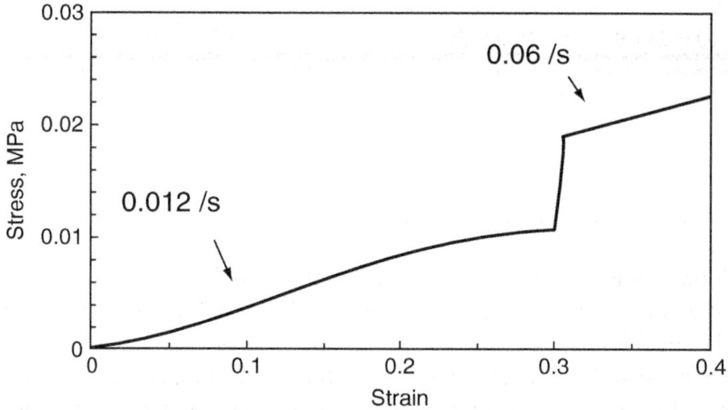

Fig. 6.19 Effect of a step change of strain rate in the case of a A356 + 20% SiC composite deformed in compression in the semisolid state. From [42]

and for a sufficient liquid fraction. During a step change of strain rate, a similar behavior to that with the nonglobular structure is observed (Fig. 6.19), because the degree of agglomeration of the globules does not evolve after the change.

Another important aspect related to the effect of shear rate is liquid segregation. This effect has been particularly observed in the case of simple compression experiments for which a free surface is present at the periphery of the specimen [32,43]. At low compression rate, liquid segregation occurs quite extensively, whereas at very high rate, no segregation is observed. The explanation for this lack of segregation is that liquid flow requires very high pressures to occur so that homogeneous deformation is more likely to operate.

Table 6.2 Values of the parameter B in the expression $\eta = A \exp(Bf_s)$ for various alloys with globular structures

Alloy	Structure	Value of B	References
Al–Si	Globular	18–19	[11]
AZ91	Globular	15	[39]
Al–Ge	Globular	15	[44]
Sn-15%Pb	Globular	20.6	[45]

Influence of the Solid Fraction

The influence of the solid fraction on the viscosity of semisolid alloys has been determined by many authors on various alloy categories. Viscosity obviously increases with increasing solid fraction. As for the rheological behavior of the alloys during solidification, exponential laws have been proposed to account for this variation ($\eta = A \exp(Bf_s)$). Table 6.2 gives the various B values obtained with globular structures.

As for the alloys during solidification, the value of B is very high, which is consistent with the agglomeration of the globules leading to a larger effective solid fraction than the actual one.

The influence of the solid fraction on the viscosity has also been determined in the case of aluminum matrix composites. Similar to the behavior during solidification, the addition of ceramic particles reduces the viscosity of the semisolid alloy at the same temperature. This result can be explained by the fact that the ceramic particles hinder agglomerate formation between the particles [41].

6.2.3 *Results in Terms of Constitutive Behavior*

Many tests carried out on partially remelted alloys impose a stress state with an important hydrostatic component. During deformation of a specimen with a free surface, the liquid is subjected to a pressure gradient, which can lead to liquid segregation. This phenomenon occurs frequently during the compression test (particularly in the case of an important friction between the specimen and the compression plates) and during direct and backward extrusion. It can happen also during a forming operation of a component. In order to predict this phenomenon, it is necessary to develop models that account for the two-phase nature of the material and to identify these models by using tests involving a triaxial stress state. Oedometric compression and drained triaxial compression have been developed with this objective. In addition, even if liquid segregation occurs mainly under compressive stress states, it is still possible for local tensile stresses to develop inside a part during forming leading to liquid sucking.

Drained compression tests allow the determination of the densification behavior of the solid phase assuming that the pressure required to expel the liquid, is negligible. Figure 6.20 shows the variation of the filtration pressure as a function

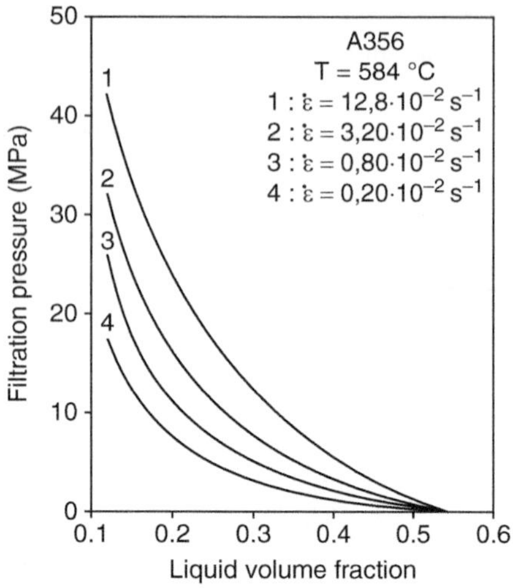

Fig. 6.20 Variation of the filtration pressure as a function of the liquid fraction remaining in the specimen during drained compression at various strain rates (from [46])

of the liquid fraction remaining in the specimen for an A356 alloy compressed at 584°C at various strain rates [46]. The filtration pressure increases with decreasing liquid fraction and increasing strain rate. These tests have demonstrated that a dendritic structure is more sensitive to densification than a globular structure at low solid fractions but the inverse is observed at high solid fraction [38]. They have also demonstrated that the liquid fraction really entrapped inside the solid globule is usually small (about 2%). In-situ X-ray microtomography carried out more recently on Al–Cu alloys has confirmed that the liquid fraction really present in the semisolid state can be quite different from that measured after solidification owing to the fact that quenching is usually not fast enough to freeze the microstructure [47].

Tensile tests have been carried out with an AZ91 magnesium alloy at high solid fractions to ensure sufficient cohesion of the solid phase. These tests have demonstrated that the tensile strength is much smaller than the compressive strength in the case of globular structures, whereas they are not so different in the case of dendritic structures [48]. This characteristic must be taken into account in the constitutive law of the semisolid alloy.

6.3 Comparison Between Partially Solidified and Partially Remelted Alloys

To the best of our knowledge, there is no detailed evaluation of the rheological behavior of the same alloy obtained by partial solidification on one hand and by solidification and subsequent partial remelting on the other hand. This comparison is obviously quite difficult to make, since it will depend on the way the alloy is partially solidified and the way it is reheated up to the temperature in the semisolid

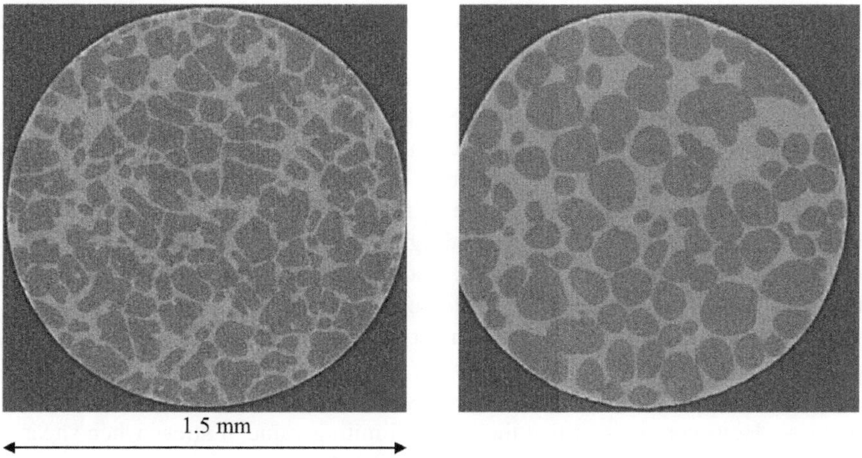

Fig. 6.21 Microstructures of an Al-15.8%Cu alloy after solidification (*left*) and after partial remelting at 555°C and holding for 80 min (*right*). These micrographs were obtained from in-situ tomography experiments carried out at ESRF Grenoble (from [50])

state for testing. In the case of conventional solidification, it has been observed that the liquid is more uniformly distributed in the interdendritic spaces than during remelting for which the liquid is more concentrated in the form of pockets. The strength of the alloy is thus smaller during partial solidification than during remelting at the same solid fraction [49]. In the case of globular structures, the situation is more complex, since strong evolution of the microstructure can take place during remelting. Solidification could lead directly to globules of the solid phase. These globules are well separated but they are not fully globular. Partial remelting leads to globularization of the microstructures but at the same time, agglomeration can occur. Figure 6.21 shows the microstructure of an Al-15.8%Cu alloy obtained during solidification with grain refinement and the same microstructure after partial remelting and holding for about 80 min in the semisolid range [50]. These microstructures were obtained by in-situ X-ray microtomography carried out at ESRF, Grenoble. It is clearly observed that the microstructure obtained during solidification is not fully globular, but most of the solid grains are isolated by a liquid film. After holding, the globules become more spherical (their shape factor decreases) but necks form in between the globules. The rheological behavior of these two types of microstructure is difficult to foresee: Globularization would decrease viscosity, whereas agglomeration would increase it.

6.4 Yield Stress

Several models used to describe the behavior of semisolid alloys include a yield stress, such as the Bingham or the Herschel–Bulkley equations (see Sect. 7.3). Basically, the yield stress is the stress below which there is no flow of the alloy.

Experiments to determine such a yield stress have been performed both on partially solidified and partially remelted alloys. Experiments were carried out by Modigell and Koke [51] by using a stress-controlled Couette-type rheometer. Increasing and decreasing linear stress ramps were used and they clearly observed no flow of the slurry below a given stress. This yield obviously depends on the material history. Other experiments have been performed by using a vane rheometer in which an alloy is progressively heated from the solid state [28]. The test consists in imposing a torque during heating and recording the temperature (i.e., the solid fraction) at which the material starts to deform. The shear stress is then calculated from this torque, which corresponds to the yield stress for this solid fraction.

Kirkwood and Ward [52] argue against the use of a yield point based on several experimental results at low shear rates [17, 23], which indicate that there is no limiting value of stress as shear rate decreases, viscosity and shear stress continuing to increase. However, the fact that there is no limiting value of stress when shear rate decreases is not similar to having a yield point. In the first type of experiments, a given nonzero stress must be applied to induce the flow of the slurry, whereas in the second, a shear rate is applied and it is observed that decreasing the shear rate leads to a continuous increase in shear stress due to the increased agglomeration of the globules. It is to be noted that this increase in shear stress with decreasing shear rate is observed if the slope of the viscosity vs. shear rate curve is smaller than -1, which is often the case in partially solidified alloys (Fig. 6.5). If this slope is greater than -1, stress will decrease with decreasing shear rate and in this case, extrapolation of the curve at zero shear rates could lead to a yield stress.

A serious objection against the existence of a yield stress is that deformation of a semisolid slurry takes place at very high homologous temperatures (T/T_{L} is close to 1 if T_{L} refers to the liquidus temperature of the alloy), so that diffusion will always take place without any yield stress. Therefore, a semisolid slurry will deform whatever the applied stress level but the resulting strain rate could be very small and even hardly detectable. This is obviously not interesting in practice so that the yield stress will be defined as the stress below which there is no detectable strain within the duration of the experiment. During injection, for which this duration is very small, the apparent yield stress could be quite large, whereas it could be much smaller during a laboratory experiment carried out at much smaller velocity. It seems therefore that the concept of yield stress is convenient to account for observations of regions described as unyielded or dead in a slurry having experienced a forming operation.

6.5 Concluding Remarks

Experimental results concerning the rheological behavior of semisolid alloys show that this behavior is very complex, since it involves many parameters which evolve during the experiment. These parameters are obviously linked both to the material (solid fraction, solid morphology, ...) and to the conditions of the experiments (shear rate, deformation mode, ...). Although a huge number of experiments have been carried out so far on many types of materials, there is still debate concerning the type of experiments to be carried out, the interpretation of the results, and their modeling. In particular, for numerical simulation of forming processes, it is necessary to carry out experiments, which reproduce at best the forming conditions. Transient situations are therefore very important since the forming operations last usually less than a second. On the contrary, steady state conditions do not prevail during forming but they are nevertheless interesting from a more fundamental point of view.

Chapter 7
Modeling of Semisolid Processing

7.1 Introductory Remarks on Modeling

Computational fluid dynamics (CFD) can be used to predict die filling. However, input parameters must be obtained from rheological experiments (see Chap. 6. In some cases, the data that has been used is from steady state experiments, where the material has been maintained at a particular shear rate for some time. In cases where the process involves taking material from rest into the die in a fraction of a second, it is difficult to see how this can be appropriate. The material changes viscosity by several orders of magnitude in that fraction of a second [23, 32]. Observations of transient rheological behavior under rapid changes in shear rate are therefore more relevant for input data for modeling.

Modeling of semisolid processing is generally continuum modeling, where the macroscopic behavior is predicted with the internal structure represented by a few internal variables. Continuum modeling for semisolid can be categorized into one-phase or two-phase and as finite difference or finite element. In one-phase modeling, a single non-Newtonian equation of state for the viscosity takes into account the interrelationship between the liquid and the solid through the way in which they behave when they are in combination. In two-phase modeling, each individual phase (the liquid matrix or the solid particles) is characterized with separate equations. The two-phase approach is more complicated, requires significantly higher computational time and much higher effort on obtaining the experimental parameters for input. It is, however, more physically realistic and does allow for important (and undesirable) phenomena, such as liquid segregation to be modeled. Liquid segregation is where liquid gathers in one part of the component in preference to another during processing. Figure 7.1 shows a classic but extreme example. The material is being forced vertically upwards into a die and has to flow around a corner to reach the end of the die. The sharp corner (the dotted line in the inset shows the die shape) causes the liquid to separate from the solid and to gather in one place. In the micrograph, the quenched liquid is dark gray in comparison with the solid. The die has not filled properly. Liquid segregation is deleterious, leading, for example, to inhomogeneous mechanical properties after solidification.

David H. Kirkwood et al., *Semi-solid Processing of Alloys*,
Springer Series in Materials Science,
DOI 10.1007/978-3-642-00706-4_7,

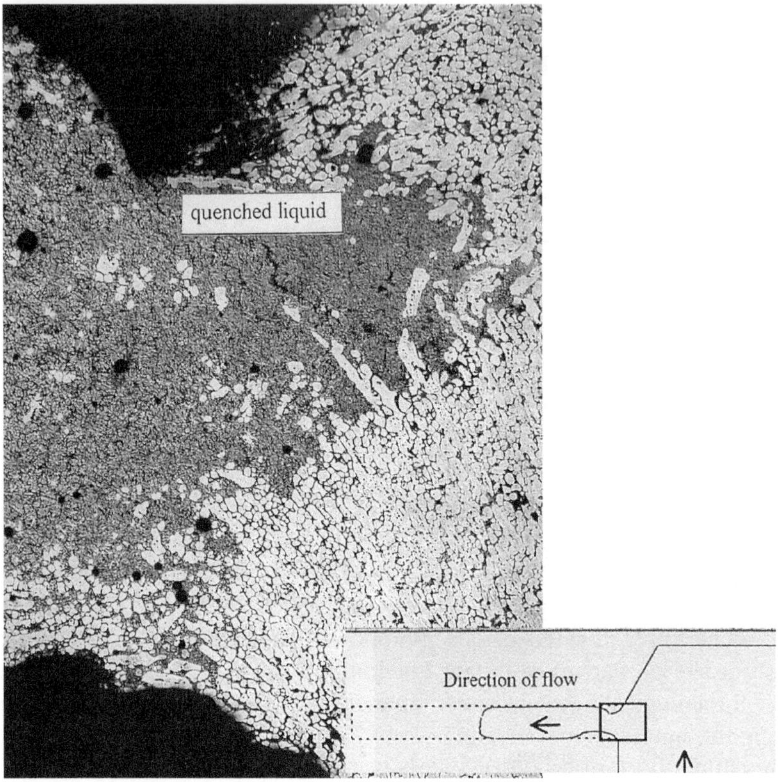

Fig. 7.1 Aluminum alloy 7075 is being forced vertically upwards into a die (*see inset*). The material has to flow around a corner and liquid segregation is occurring at that corner [53], partly because the temperature conditions are inappropriate. The *rectangular outlined portion* is the area shown in the micrograph

Continuum modeling generally requires discretization for analysis. The major methods used in semisolid modeling are the finite difference method and the finite element method. In the finite difference method, approximations for the derivatives at a grid point are chosen. In contrast, in the finite element method, weighting functions are used for the grid nodes. A detailed discussion of these discretization methods is given in [54, 55].

Modeling of semisolid processing, by its very nature, sits at the interface between fluids modeling and solids modeling. The CFD approaches originate in fluids modeling but in semisolid processing must bring in the thixotropic behavior. For higher solid fractions, solid mechanics approaches may be more appropriate. In effect, if fractions of solid are nearer 0.8 or 0.9, the process may be more akin to soft forging. Processing in this regime may not be what we might think of as classical thixo- or rheo-forming. Examining the microstructure after forming will tend to reveal deformed solid spheroids (see Fig. 7.2) rather than the round spheroids, which would

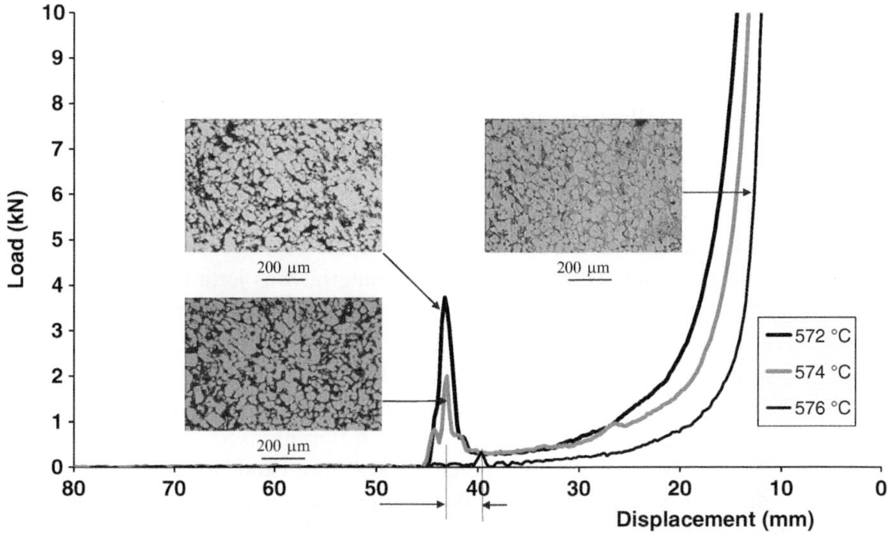

Fig. 7.2 Load versus displacement for the rapid compression of a billet of A357 aluminum alloy in a thixoformer [56], illustrating that at relatively high fraction solid (572°C), the spheroids in the microstructure are deformed. At relatively low fraction solid (576°C), the spheroids are undeformed and the resistance to deformation is low. The closeness of these temperatures shows how sensitive the process is to temperature. The microstructures are taken from the edge of the billet after compression has been completed

be expected to exist throughout the solidified component in classical thixoforming. It is helpful to include consideration of such solid approaches here. In addition, there have been recent developments of micromechanical continuum models (termed here "micromodeling"), which involve the concept of a "coated inclusion" and sit at the interface between fluids modeling and solids modeling.

Previous reviews can be found in [57–60]. Atkinson [59] provides an extensive survey of modeling work in the literature and the validation of the results, including summarizing the approaches taken by different modelers. For details of the mathematical basis, it will be helpful to consult [60]. Here, we take an overview with relatively little mathematical detail. Section 7.2 will highlight the utility of modeling, Sect. 7.3 will introduce the use of the structural parameter λ for modeling and outline the model of Brown and coworkers, which is extensively drawn on by other researchers. Section 7.4 briefly highlights the issue about the use of yield points in modeling. Section 7.5 then covers finite difference modeling and Sect. 7.6 finite element, with one- and two-phase modeling covered in both cases. Models akin to forging models, essentially those where the time dependence of viscosity is not treated, are covered in Sect. 7.7. Micromodeling is then discussed in Sect. 7.8. Some concluding remarks are then given.

7.2 Utility of Modeling

Die design rules from die casting are not transferable to thixoforming and related processes. Figure 7.3 illustrates this problem with a generic demonstrator consisting of a round plate with three bolt holes and a central boss [61]. When preliminary trials were carried out, there was difficulty in filling the die. Partial filling experiments were carried out, which showed that the design of the die, particularly the in-gate, which was narrow and "inward directed" as for die-casting, was leading to some jet flow across the cavity (Fig. 7.3a); i.e., the material was jetting to the end of the cavity and then flowing back on itself. Defects were occurring behind the central boss and on either side just after the entrance. The flow was contrary to the smooth progressive filling, which is the aim in thixoforming.

FLOW3D, which is a CFD program from FLOWSCIENCE Inc., was used to model the flow in the die (Fig. 7.3b), trying out different viscosities in order to find the range in which the experimental behavior was mimicked. The work did not take into account either heat transfer in the die or friction at the die surface. Even so reasonable agreement was found. In addition, the model, in the version used, did not allow for thixotropic behavior as such (i.e., it assumed the fluid had a constant viscosity independent of the shear rate and time, which in practice is not the case). The design of the die was changed as a result of this "simple" modeling and this led to improved filling (Fig. 7.3c). The in-gate was widened and deepened which reduced jet flow. In addition, the bolt holes were made ellipsoidal, which reduced disruption

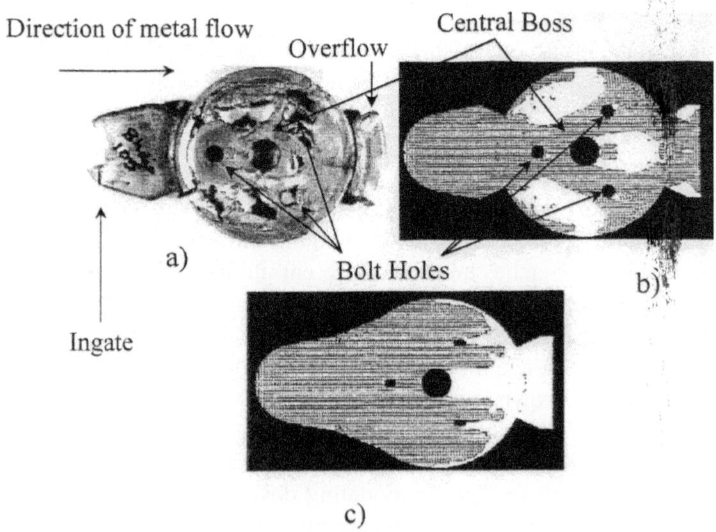

Fig. 7.3 Numerical simulation of die filling [61]. (**a**) Partial filling of die. (**b**) Modeling simulation of (**a**) where *white* corresponds to *dark* on (**a**). (**c**) Modeling simulation with improved die design showing smoother filling

to the flow path. In practice, the latter would not be possible because the shape is dictated by the requirements of the component. However, the wide, deep, smoothly contoured character of the in-gate is now widely recognized to be a requirement for thixoforming. This example illustrates the real potential commercial benefit, which can be obtained from better understanding of flow of semisolid material in dies. The use of modeling can reduce the need for extensive trials of die design, ram speed, dwell time, and pressure.

7.3 Use of the Structural Parameter λ for Modeling: Model of Brown and Coworkers

As described earlier, thixotropy is associated with the formation of bonds between spheroids (and agglomerates of spheroids), when the material is at rest or being sheared at low shear rate. When the shear rate is suddenly increased, some of these bonds will be broken and the viscosity decreases. The extent to which a continuous structure of bonded spheroids is in existence is represented by what is called the structural parameter λ (sometimes called the "single internal variable" in the modeling of semisolid processing). A fully built-up structure is usually represented by $\lambda = 1$ and a completely broken-down structure, with all the spheroids separate from each other, by $\lambda = 0$ [62]. At a particular shear rate, after enough time has elapsed, the processes of break-down and build-up will be in dynamic equilibrium so that a particular value of λ is characteristic of that shear rate. Thixotropy is then introduced mathematically via the time derivative of the structural parameter $\mathrm{d}\lambda/\mathrm{d}t$. This is the balance of the build-up and break-down terms. In the simplest theories, these are only controlled by the shear rate and the current level of the structure parameter λ. The driving force for build-up is controlled by the difference of the structure from its maximum value, i.e., $(1 - \lambda)$, raised to a power b. The rate of break-down due to shearing can be given by the product of the current level of the structure λ and the shear rate $\dot{\gamma}$ raised to another power d. So,

$$\frac{\mathrm{d}\lambda}{\mathrm{d}t} = a\,(1 - \lambda)^b - c\lambda\dot{\gamma}^d, \tag{7.1}$$

where a and c are constants for any one system. The system is breaking down towards equilibrium if the overall value of $\mathrm{d}\lambda/\mathrm{d}t$ is negative and building up if it is positive. The Moore model [63] is a simplified version of (7.1) with b and d set to 1. Cheng and Evans [64] set b to 1 but allowed d to take different constant values.

The next step in any model is to relate the structure λ to the stress τ or viscosity η in a flow equation. This has been done in a variety of ways including the simple Bingham model

$$\tau = \tau_y + k\dot{\gamma}, \tag{7.2}$$

where τ_y is a yield stress, and the Herschel–Bulkley model where the behavior is nonlinear after yield

$$\tau = \tau_y + k\dot{\gamma}^n. \tag{7.3}$$

Brown and coworkers [21,65,66] presented a constitutive model based on the single internal variable concept. The two key equations are

$$\tau = A(\lambda)\frac{(c/c_{\max})^{1/3}}{1-(c/c_{\max})^{1/3}}\eta_f\dot{\gamma} + (n+1)\,C(T)\,\lambda f_s \eta_f^{n+1}\dot{\gamma}^n, \tag{7.4}$$

$$\frac{d\lambda}{dt} = H(T, f_s)(1-\lambda) - G(T, f_s)\,\lambda\dot{\gamma}^n. \tag{7.5}$$

In (7.4), $A(\lambda)$ is a hydrodynamic coefficient, which is a function of λ, c is the effective volume packing fraction solid and $c_{\max}$ the maximum value of that, η_f is the viscosity of the fluid, n is the shear rate exponent, $C(T)$ an exponential function of temperature and f_s the fraction of solid, which is also temperature dependent. In (7.5), H is the agglomeration function and G is the disagglomeration function, both of which are dependent on the temperature and the fraction solid. Brown et al. assume that flow resistance is due to hydrodynamic flow of agglomerates (the first term on the right in (7.4)) and deformation of solid particles within the agglomerates (the second term on the right). The hydrodynamic coefficient $A(\lambda)$ depends on the size, distribution, and morphology of the particle agglomerates. The deformation resistance term arises because of the energy dissipated in the plastically deforming particle–particle bonds. Overall, the model of Brown et al. predicts an increase in the deformation resistance with the solid fraction and this becomes rapid between 0.5 and 0.6 f_s. It is not valid beyond about 0.6 f_s.

The constitutive equations due to Brown et al. [21, 65, 66] have been used by a number of workers both for finite difference modeling and finite element [67–70], sometimes with the introduction of a yield stress into (7.4) [67].

7.4 Use of a Yield Stress in Models

There are very mixed views about whether a yield point should be used in modeling semisolid processing. A number of the models described below include a yield stress (e.g., those based on Bingham or Herschel–Bulkley equations – see (7.2) and (7.3)). Kirkwood and Ward argue against the use of a yield point [52] (see Sect. 6.4), defining a yield point as the stress approached, as the flow is reduced. This is distinct from the peak seen in compression tests, which is associated with the initial structural breakdown of the solid "skeleton." This is of a different magnitude from that proposed in Bingham or Herschel–Bulkley modeling.

7.5 Finite Difference Modeling

7.5.1 *One-phase Finite Difference Modeling*

Ilegbusi and Brown [67] used the Brown et al. model (see Sect. 7.3) with a yield stress to examine flow into a chisel-shaped cavity. A solid shell formed at the mould wall leading to "jetting" in the center of the cavity and hence showing the importance of heating the die and the heat transfer coefficient.

A commercial code called FLOW3D has had a thixotropic module incorporated. The mathematical approach is outlined in [71,72]. The two key equations are given as (7.6) and (7.7). A transport equation (7.6) is used for the viscosity η rather than for the structural parameter λ (as in the Brown et al. model), because this is convenient for computational fluid dynamics, which requires a value for η.

$$\frac{\partial \eta}{\partial t} + \underline{u} \cdot \nabla \eta = \omega \left(\eta_{\mathrm{e}} - \eta \right), \tag{7.6}$$

$$\frac{\partial \lambda}{\partial t} + \underline{u} \cdot \nabla \lambda = b_1 \left(1 - \lambda \right) + b_2 \lambda \dot{\gamma}. \tag{7.7}$$

$\underline{u}$ is the velocity vector, ω is the inverse of the relaxation time, η_{e} is the steady state equilibrium viscosity and b_1, b_2 are constants. The second term on the left of (7.6) is the advection term and the term on the right is the relaxation term. The relaxation depends on the steady state viscosity η_{e} and the relaxation time $1/\omega$. Both of these variables depend on shear rate and solid fraction. The model does not include yield stress, wall slip, or elastic or plastic behavior at high solid fractions. It is most relevant where fractions solid are less than about 0.7. Agglomeration and disagglomeration are represented on the right hand side in (7.7). Here disagglomeration is dependent on $\dot{\gamma}$, and not $\dot{\gamma}^n$ as in (7.5) in the model of Brown and his coworkers. The model has been used to predict hysteresis curves for semisolid Sn15%Pb with a reasonable degree of accuracy [71]. Equations (7.6) and (7.7) are in fact equivalent, but both are given here for clarity in explanation.

The rate coefficient ω used by Barkhudarov et al. [72] has a linear dependence on shear rate $\dot{\gamma}$. This can be compared with that of Kirkwood and Ward [52], which has a linear dependence on $\dot{\gamma}^m$. If $m = 1$, the two equations for ω are equivalent.

Barkudarov et al. present their model in slightly different terms [72], which is helpful for developing understanding. They used equation

$$\frac{\partial \eta}{\partial t} + \underline{u} \cdot \nabla \eta = \alpha \mathrm{Min} \left(\eta_{\mathrm{e}} - \eta, 0 \right) + \beta \mathrm{Max} \left(\eta_{\mathrm{e}} - \eta, 0 \right). \tag{7.8}$$

$\mathrm{Min}\left(\eta_{\mathrm{e}} - \eta, 0 \right)$ means take the lower value of $\eta_{\mathrm{e}} - \eta$ and 0 and $\mathrm{Max}\left(\eta_{\mathrm{e}} - \eta, 0 \right)$ means take the higher of the two values. If $\eta_{\mathrm{e}} - \eta < 0$, then the right hand side is equal to $\alpha \left(\eta_{\mathrm{e}} - \eta \right)$ and if $\eta_{\mathrm{e}} - \eta > 0$, then the right hand side is equal to $\beta \left(\eta_{\mathrm{e}} - \eta \right)$. What this means is that if the local viscosity is greater than the equilibrium viscosity η_{e}, then the local viscosity is driven towards η_{e} at the thinning rate α. If

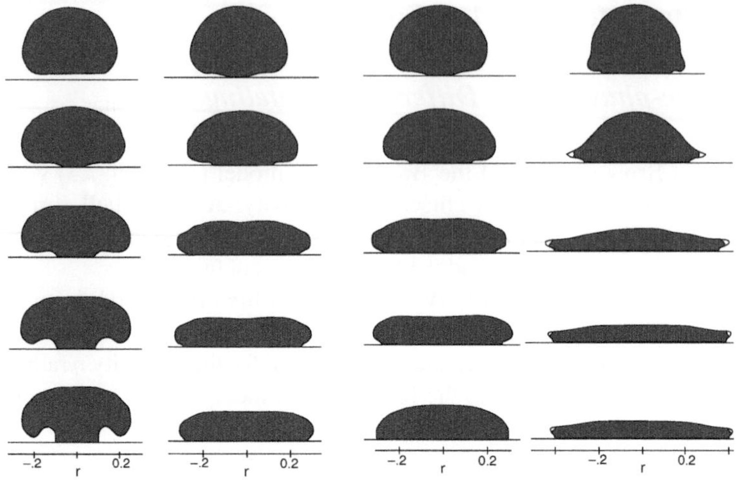

Fig. 7.4 Comparison of the effect of thixotropic relaxation time when a droplet of Sn15wt%Pb is allowed to fall from rest onto a plate located 6 cm below [72]. The droplet is assumed to be at 197°C with approximately 47% solid fraction. Heat transfer to the plate is taken into account. Relaxation times (in ms) from *top to bottom* are 1, 2, 5, 10, and 50. The *first column on the left* has a zero rate, i.e., constant viscosity. The *second* has thinning and thickening rates of 1,000 s^{-1}. The *third column* has a 1,000 s^{-1} thinning rate and a 0.0001 thickening rate. The *last column* has infinitely fast rates

the local viscosity is less than the equilibrium viscosity η_e, then the local viscosity is driven towards η_e at the thickening rate β. α and β will be dependent on the shear rate and it is known that shear thinning tends to be much faster than shear thickening. Figure 7.4 [72] shows the results for a droplet of Sn15wt%Pb impacting on a flat plate having been dropped from rest 6 cm above it. The columns show the behavior for different relaxation times. Infinitely fast relaxation gives very fast "spreading" (see the extreme right hand column in the Fig. 7.4). Zero relaxation leads to a mushroom-type shape (see the extreme left hand column in the figure). Figure 7.5 [72] shows detail of the impact process with surface tension dominating initially, but then gravity taking over in governing the behavior.

A number of workers have used the FLOW3D thixotropic module [73–77] for modeling semisolid processing. Modigell and Koke [73] used the Herschel–Bulkley model (7.3) with the yield stress τ_y dependent on the fraction of solid. They modified the nature of the structural parameter making it into one, which varies between 0 for fully broken-down and infinity for fully built-up (rather than between 0 and 1). The time evolution of the structural parameter is described with first-order reaction kinetics. The model fits shear rate experiments with Sn15%Pb, where the shear rate undergoes step changes, quite well. The parameters are obtained by iterative adjustment. Figure 7.6 shows simulation of die fill in a cavity with a cylindrical obstacle, contrasting the behavior with Newtonian and thixotropic fluids. With Newtonian fluid, the flow jets past the obstacle and hits the end of the cavity before beginning to flow back on itself. The thixotropic fluid is flowing smoothly past the obstacle

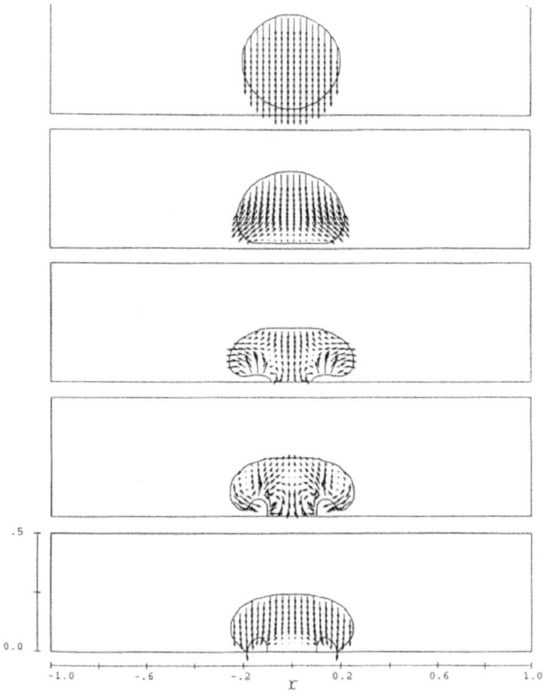

Fig. 7.5 Impact of a viscous drop with the same material as for Fig. 6.5 [72]. Times in ms from *top to bottom* are 0, 1, 5, 50, 100. Surface tension is dominating the flow in the 5–50 ms period; gravity is dominant by 100 ms

and the flow front may well meet and remerge with the material flowing on the left hand side of the obstacle (but not shown here because of symmetry).

Modigell and Koke [51] found that above a critical inlet velocity the filling was not laminar any more. They found that they could represent the transition between laminar and turbulent filling (in the sense of a smooth flow front and one that is starting to break up) reasonably well.

The FLOW3D thixotropic module can be used to model rheometry experiments (with a concentric cylinder viscometer, for example) as well as die fill. It is important to ensure that possible sources of error in modeled shear stress curves for rheometry experiments are understood. When the inner cylinder of a viscometer (the "bob") is accelerated, a finite time is required for the momentum to propagate across the gap between the cylinders. This diffusion produces a stress peak, even in Newtonian fluids. The propagation time (which is equivalent to the width of the shear stress peak) can be calculated approximately, based on the momentum diffusion equation in a plane sheet [74],

$$t = 0.45\frac{x^2}{\nu}, \tag{7.9}$$

where x is the gap width between the rotating cylinder and the cup and ν is the kinematic viscosity, i.e., η/ρ, where η is the fluid viscosity and ρ the density. For a fluid of 1 Pa.s viscosity, the time is $\sim$1 ms, which is too short to register on most (if not all) data collectors in rheometers. FLOW3D gives the option of either an

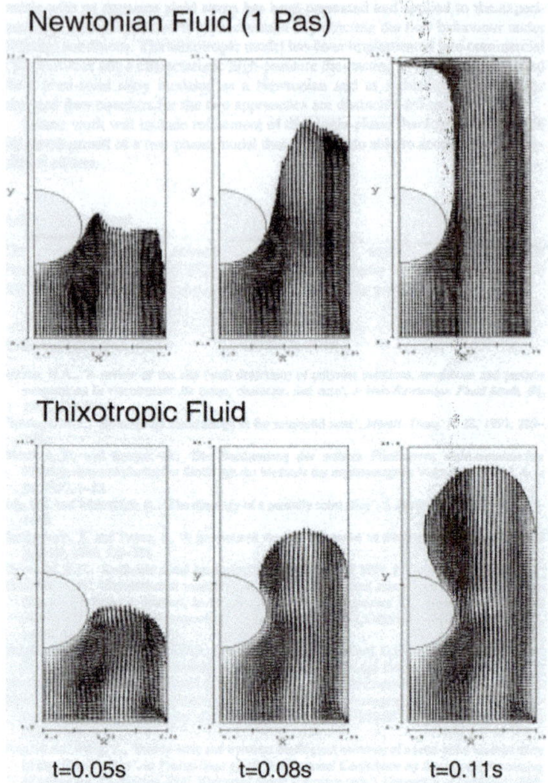

Fig. 7.6 Comparison between simulation of flow into a cavity with a round obstacle assuming Newtonian behavior and assuming thixotropic behavior [73]. The direction of flow is upwards

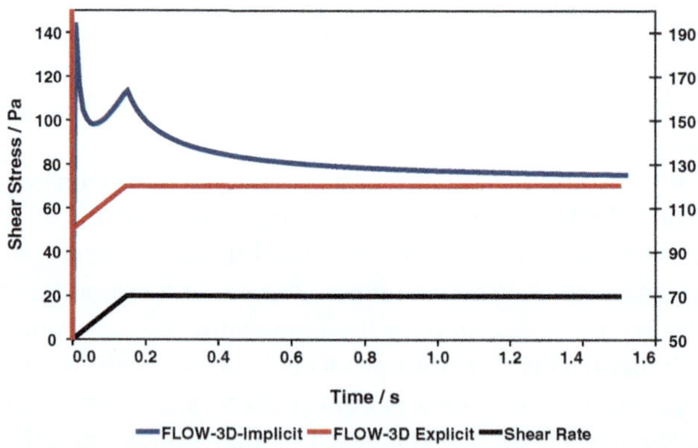

Fig. 7.7 Models of shear rate jump from 50 to 70 s^{-1} in a 1 Pa.s Newtonian fluid in a rotational viscometer [75]. The implicit solver overestimates the width of the momentum diffusion peak

implicit solution or an explicit solution to the Navier–Stokes equation. The shear stress peak height has been modeled with both implicit and explicit solvers [75]. Figure 7.7 shows the result for a linear jump in shear rate from 50 to 70 s^{-1} (with the jump taking 150 ms) in a viscometer. The fluid is Newtonian with a viscosity of 1 Pa.s. The implicit solver overestimates the length of the momentum diffusion peak, which should last only 1 ms, and its height. The explicit solver accurately reproduces Newtonian behavior, with a very sharp initial momentum diffusion peak. Experiments with a 1 Pa.s Newtonian silicone oil confirmed the explicit prediction.

Only by studying a Newtonian fluid, it was possible to see the overestimation of the momentum diffusion time and the stress produced by the implicit solver. For a shear rate jump in tin–lead (the standard model thixotropic alloy), the solver would be unable to reproduce the acceleration part because the results would combine effects due to thixotropic breakdown with the (overestimated) momentum term. The overestimation will have a small effect on die filling flow patterns, but may not be accurate for predicting the pressure distribution in the die. However, the problem with the explicit solver is that, at high thixoforming viscosities, the computing times are inordinately long. Ward et al. [76] showed that there was a need for a new solver in FLOW3D to cope with the fact that viscosities change over many orders of magnitude in short distances and times. A new alternating direction implicit (ADI) solver was then incorporated. Figure 7.8 shows a comparison of the modeling for a shear rate jump in a viscometer with experimental results (three repeats of the same experiment) and a one-dimensional spread sheet calculation. The fit is reasonable. The potential problems with implicit solvers are not mentioned elsewhere in the literature and it is not clear whether other researchers have tested their modeling for such artifacts.

Thixoforging (as opposed to thixoforming or thixocasting) involves inserting the semisolid (with a relatively high solid fraction) directly between open dies and then bringing the parts of the die together with a ram. Messmer [77] has used FLOW3D to simulate this process, which involves simulating moving die halves. The apparent

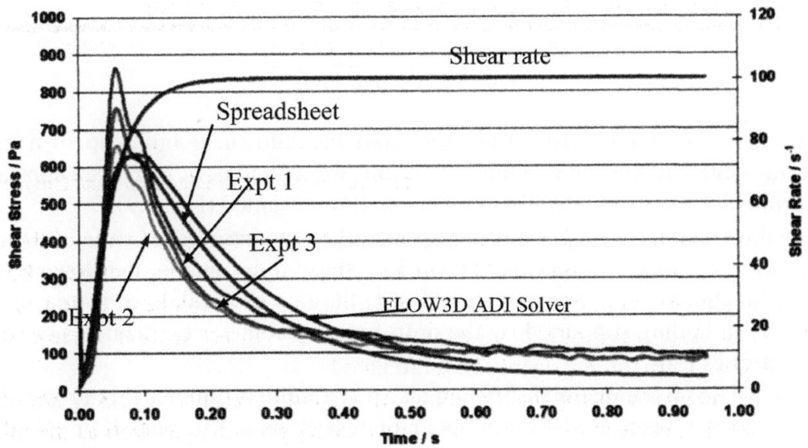

Fig. 7.8 Shear rate jump from 1 to 100 s^{-1} in SnPb alloy ($f_s = 0.36$) showing repeats of the same experiment and modeled fits using a spreadsheet and FLOW-3D predictions with a new ADI solver, which was introduced to address the challenges posed by fluids where the viscosity and shear rate change by many orders of magnitude in short times and spaces [76]

viscosity depends on fraction solid f_s (calculated using the Scheil equation), shear rate $\dot{\gamma}$, and time t The approach is essentially based on that of Barkhudarov et al. [71] with two key equations, one for the equilibrium viscosity η_e and other for the time dependent thixotropic effects:

$$\eta_e = A \exp\left(Bf_s\right) \dot{\gamma}^m, \tag{7.10}$$

$$\frac{d\eta}{dt} = \beta\left(\eta_e - \eta\right). \tag{7.11}$$

The viscosity parameters were obtained by fitting the simulation to experimental results with A356 aluminum alloy. The forming force measured at the end of the stroke corresponds well with the simulated force. The early part of the stroke is not well simulated. This is thought to be associated with the use of only one thinning rate, when in practice there may be a series of thinning rates depending on the shear rate. The initial thinning rate is thought to be much higher than that in the final stages. This is consistent with thinking by Ward et al. [76] and with the proposal by Quaak [3] (see Sect. 2.1.1) that at least two different relaxation processes are operating, with different characteristic relaxation times.

FLOW3D has recently been used to analyze the behavior in a drop forge viscometer [30–32] (see section "Parallel Plate Compression"), where a small billet is compressed by a falling steel plate and the compression followed in real time with a video camera [78]. The analysis depends on a numerical model for the coupled motion of a rigid body (the steel plate) and a fluid. It is analogous to that for a plunger, which is pushing a fluid into a die and where the plunger motion is affected by the build-up of pressure in the fluid as it flows into the die.

Kirkwood and Ward [52] use FLOW3D to simulate rapid compression tests between parallel plates. They use a rate equation for the thixotropic breakdown,

$$\tau = \frac{1}{(a + b\dot{\gamma}^m)}, \tag{7.12}$$

where τ is the characteristic time for rapid breakdown or build up to a pseudosteady state structure (not in this case shear stress), and a, b, and m are constants. Experimental results with Al–Si alloy are well represented (Fig. 7.9).

As the soaking time prior to compression is increased, the structure breaks down to a significant extent so that, in low shear rate regions, the flow is close to Newtonian. At shear rates above 100 s^{-1}, the rate of breakdown increases dramatically, a finding supported by the drop forge viscometer experiments of Yurko and Flemings [30–32].

MAGMASoft commercial software also has a finite difference thixotropic module, which has been used by a number of researchers. Kim and Kang [79] and Seo and Kang [80] have carried out simulations for a thixotropic fluid obeying the Ostwald-de-Waele power law (i.e., (7.3) but without the yield stress τ_y) with the exponent n taking various values. There is reasonable agreement between the

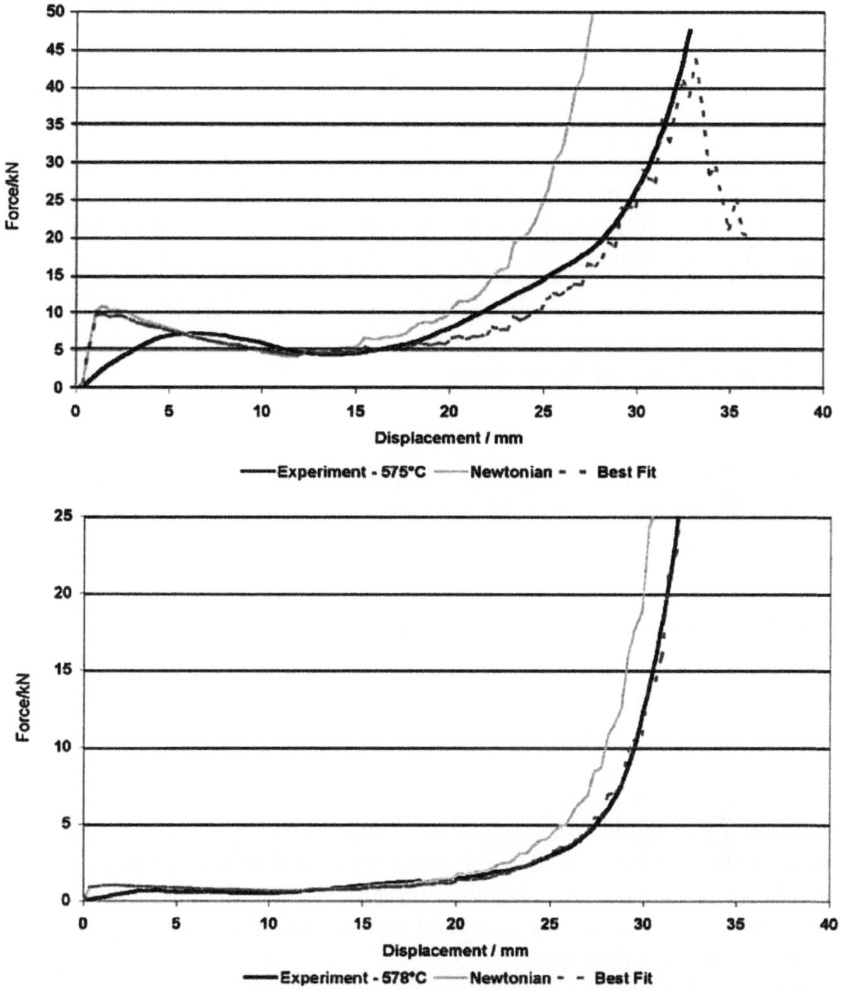

Fig. 7.9 Experimental and modeled results (with FLOW3D) for compression tests on A356 aluminum alloy at two different temperatures [52]. For the higher temperature (578°C), there is hardly any initial peak at the beginning of the curve

simulation and the results of partial filling experiments for components, which do not involve parting of flow fronts and rewelding beyond obstacles or very big changes in section thickness.

Itamura et al. [81] have used a code called Adstefan to compare simulation of die casting, squeeze casting and rheocasting for both metal flow and solidification. Few details are given. The results indicate that there would be fewer shrinkage defects in rheocasting than with the other processes.

7.5.2 *Two-phase Finite Difference Modeling*

Two-phase finite difference models are rare. In Ilegbusi et al. [68], single phase equations are solved for the whole filling phase. For a given number of particles, the trajectories are computed. It is assumed that the particles will "disappear" when they hit a wall or are trapped in a recirculation zone. Comparing the number of particles at a given distance from the inlet with the total number of injected particles gives a measure of segregation.

7.6 Finite Element Modeling

As for finite difference modeling, both one-phase and two-phase treatments are available. A variety of commercial and other codes are used. The discussion below is grouped according to author(s) for clarity.

7.6.1 *One-phase Finite Element Modeling*

Zavaliangos and Lawley

Zavaliangos and Lawley [69] use identical equations to Ilegbusi and Brown [67] (based on the model of Brown et al. (see Sect. 3.3)), but without a yield stress. For fractions of solid less than about 0.5, for Sn15%Pb, it is predicted that a free standing billet will collapse. The thixoforming of a simple shape is simulated (in a thixoforging type operation) and the disagglomeration process predicted through the distribution of values of the structural parameter λ (Fig. 7.10). No experimental validation is given in the paper.

Backer

Backer [70] programmed various rheological models into software called WRAFTS. The models included a Newtonian, a shear-thinning model, and an internal variable model, but where, in this case, the variable λ is viewed as a chemical concentration term and its variation with space and time is described by a convective transport equation. The results are shown in Fig. 7.11 for a complex die. With the Newtonian rheological model, the flow front breaks up. There are a number of locations behind cores that remain unfilled as the liquid flows past them. These would then tend to be regions where defects exist in the final component. For the shear thinning model (Fig. 7.11b), the fluid fills behind cores. With the internal variable model, flow tends to occur from the runner at the side of the cavity, rather than that at the bottom. This occurs because the structural parameter λ is reduced as the mixture flows through

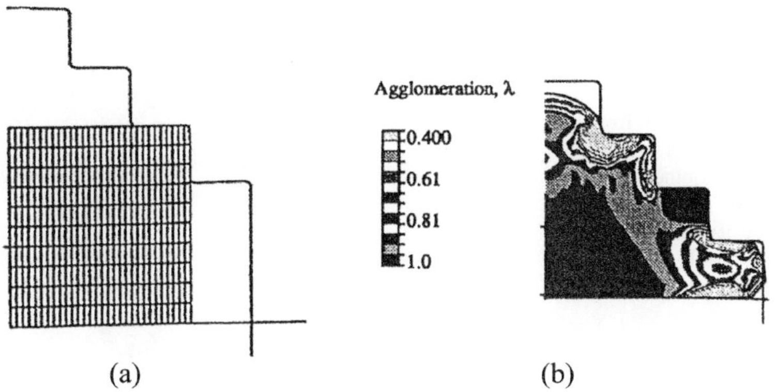

Fig. 7.10 Die filling with semisolid Sn15%Pb (f_s = 0.55) [69]. (**a**) Stepped die shape with the undeformed finite element mesh in place ready for deformation. (**b**) Part way through the deformation before die filling is completed showing the distribution of the structural parameter λ, with $\lambda = 1$ representing a structure, which is still fully agglomerated

the runner system from $\lambda = 1$ in the shot sleeve to $\lambda = 0.2$ in the side runner. The value in the bottom runner is 0.4. Thus, the material in the side runner is less viscous and can flow into the cavity more easily. No experimental validation is presented in the paper.

Alexandrou, Burgos and Coworkers

Alexandrou, Burgos and coworkers have published several papers where their approach has gradually been developed [82–88]. They use the commercial code PAMCASTSIMULOR, with a continuous Bingham law based on that introduced by Papanastasiou [89] to avoid the discontinuity in the yield stress. In simple shear terms,

$$\tau = \tau_y \left(1 - \exp\left(-m\dot{\gamma}\right)\right). \tag{7.13}$$

The yield stress τ_y is a function of the fraction solid f_s. When the local stress is higher than τ_y, the slurry behaves as a non-Newtonian fluid with m controlling the exponential rise in the stress at small strain rates $\dot{\gamma}$.

Figure 7.12 shows a comparison of the filling when the fluid is treated as Newtonian and that when it is a Bingham fluid (i.e., the power law exponent n is 1; the fluid is neither shear thinning nor shear thickening, but there is a yield stress). As in Fig. 7.11, the Newtonian fluid tends to jet across the cavity and then flow back on itself, potentially generating defects, particularly behind the core. With the Bingham fluid, the flow fronts meet beyond the core and can merge again. In addition, any oxide films on the surface of the flow front will tend to be incorporated in the component in the Newtonian case, whereas in the Bingham case, the velocity vectors beyond the core are pointing forwards and the oxide will be carried ahead of the flow into overflows, which can be designed in at the end of the flow path.

Fig. 7.11 The fluid front for die filling with different viscosity models [70]. (**a**) Newtonian showing break-up of the flow front, (**b**) shear-thinning viscosity with a smoother flow front profile, (**c**) internal variable viscosity again with a smoother profile but with some significant differences from (**b**)

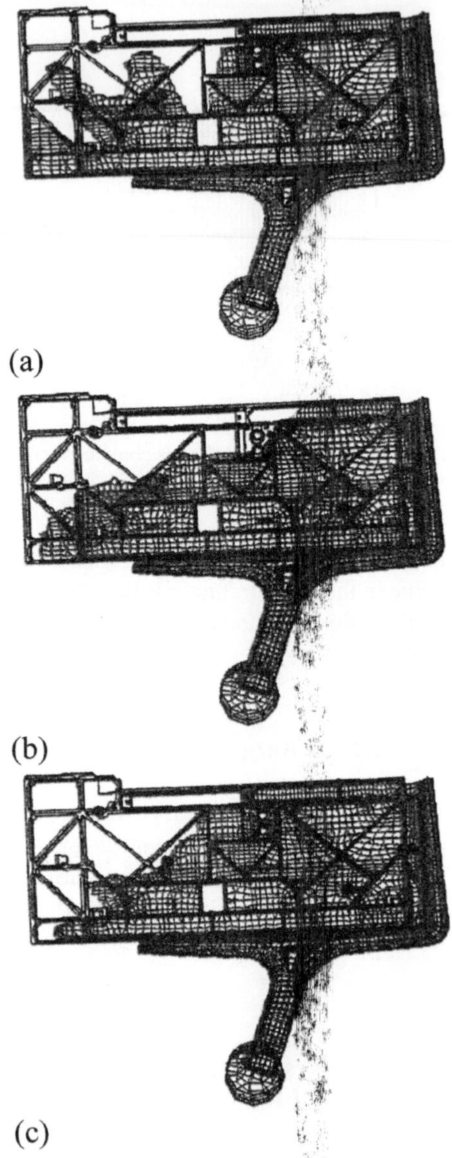

Burgos and Alexandrou [83] simulated a Herschel–Bulkley fluid (i.e., a fluid which obeys a power law dependence of viscosity on shear rate and also has a yield stress) in a sudden three-dimensional square expansion. During the evolution of the flow, two core regions of unyielded material, and dead zones at the corners, are formed.

Pan et al. [84] examine the relative importance of the inertial, viscous, and yield stress effects on the filling profile in a two-dimensional cavity with a Bingham fluid.

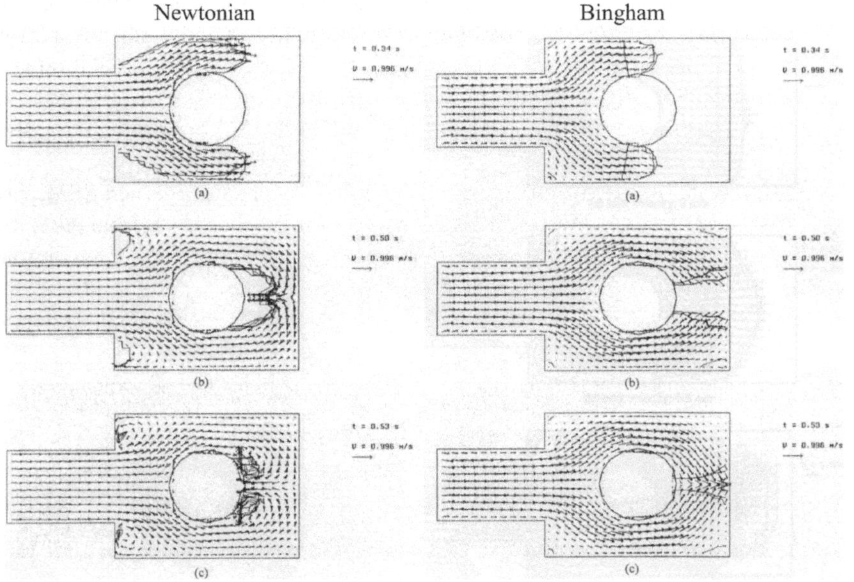

Fig. 7.12 Comparison between Newtonian (*on the left*) and Bingham (*on the right*) filling behavior for a three-dimensional cavity with a cylindrical obstacle [82]

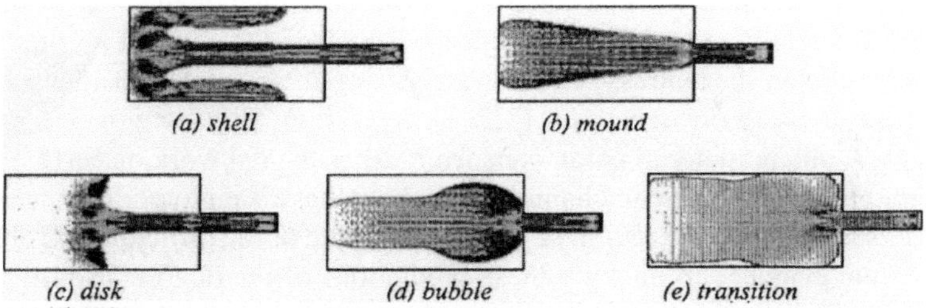

Fig. 7.13 Flow patterns found by modeling for a Bingham fluid in a two-dimensional cavity [84]

Five different flow patterns are identified (Fig. 7.13): "shell" for large Reynolds numbers, but small Bingham numbers; "mound" for low Reynolds and Bingham numbers; "bubble" for larger Bingham numbers; "disk," which occurs between shell and bubble filling and "transition."

The rather powerful approach that emerges here is the capacity to plot maps, such as that in Fig. 7.14, which indicate the regimes where different types of filling are likely to be found. The map plots the Bingham number $\tau_y H/\eta V$ versus the Reynolds number $\rho V H/\eta$ where H and V are characteristic length and velocity scales. The results indicate the importance of the yield stress relative to the inertia forces. Transition flow occupies a narrow region between the disk and the bubble patterns. This region may be prone to instabilities, and hence to defects, as the flow initially starts as a disk and then switches to bubble filling.

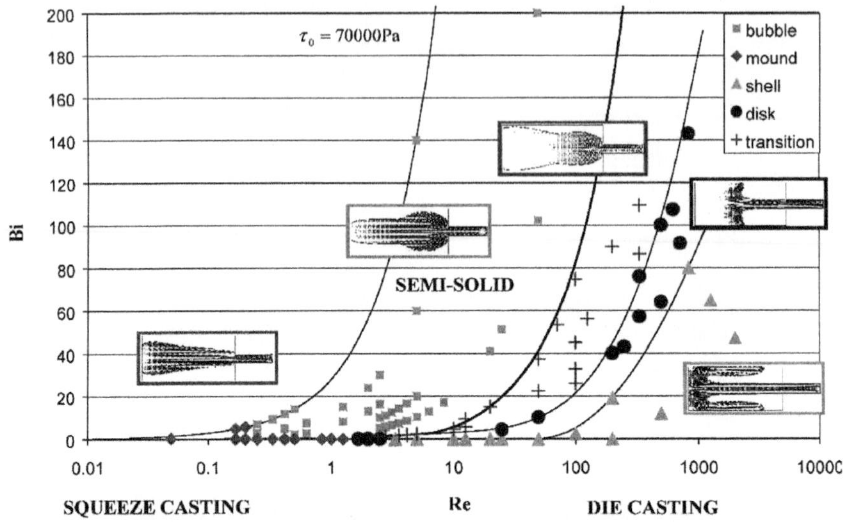

Fig. 7.14 Map showing the flow patterns identified in Fig. 34, plotting Bingham number *Bi* versus Reynolds number *Re* [84]

The approach is expanded by Burgos et al. [85] to include the effect of the evolution of the microstructure via an equation for $d\lambda/dt$, which includes an exponential factor for disagglomeration. This dependence is included to account for the fact that experimentally the shear stress evolution for shear rate step-up experiments is faster than that for the step-down case [3, 11, 23] (see section "Influence of Shear Rate"). Burgos et al. obtain material parameters from experimental work on Sn15%Pb. For flow through a straight square channel, they show that disagglomeration is small in the corners and in the core region of the channel. One of the difficulties is obtaining experimental evidence to back up these predictions of the degree of disagglomeration. One would have to assume that agglomeration can be diagnosed from the number and width of "bridges" between solid globules. However, there are stereological difficulties in identifying whether globules, which are not joined in a two-dimensional section, may be joined in three dimensions [90]. In addition, it would have to be assumed that during a quench no further bonds formed and it is already known that significantly more solid can be laid down on solid globules that exist at the semisolid temperature during a quench at a typical rate [91] (see also Sect. 6.2.3).

Alexandrou et al. [86] and Florides et al. [87] analyzed the flow and shape evolution during the compression of a finite amount of a Bingham plastic under both constant load and constant velocity. Results for various Reynolds and Bingham numbers are presented. There is unyielded material at the top and the bottom of the compressed cylinder, in stagnant layers. These results relate to the compression tests that can be used to obtain rheology parameters for thixotropic materials (see section "Parallel Plate Compression"). However, these particular simulations do not take account of the evolution of the internal structure, i.e., the time and shear rate dependence of the viscosity. There is no detailed comparison with experimental results.

Fig. 7.15 Flow instability of the "toothpaste" type in semisolid processing. The metal is filling from the right to the left [88]

Building on the identification of flow regimes prone to instabilities in [84], Alexandrou et al. [88] have analyzed two-dimensional jets of Bingham and Herschel–Bulkley fluids impacting on a vertical surface at a distance from the die entrance. Shell, disk, and mound patterns, and most transition cases, are all stable, but a bubble pattern gives an unstable jet. Instabilities are the result of finite yield stress and the way the yielded and unyielded regions interact. A common defect in semisolid processing is the "toothpaste effect" (Fig. 7.15) and this analysis gives some insight into the reason for this defect.

Alexandrou and Georgiou [92] have recently modeled the early breakdown of semisolid suspensions in the gap of a cylindrical rheometer in order to provide further insight into the short-term material transients, which are most relevant to the injection of semisolid into a die. A Herschel–Bulkley fluid model is used with the yield stress decomposed into two parts. The first relates to the breakdown during the early stages of deformation (the breakdown of the solid skeleton is referred by Kirkwood and Ward in [52]- see Sect. 6.4). The second part survives the shearing for a long time. The first component is assumed to be a linear function of the coherency parameter. The second component is relevant to steady state experiments.

Orgéas and Coworkers

Orgéas et al. [93] have used the power law cut-off (PLCO) model in Procast commercial software. The material is assumed to be a viscous, isotropic, incompressible fluid. In the Orgéas et al. work, there is one cut-off value $\dot{\gamma}_0$ and this is determined by geometry. Shear thinning only occurs if the shear rate $\dot{\gamma}$ exceeds the cut-off value. Essentially, these researchers are assuming that *agglomeration* probably does not take place over the very short injection times characteristic of thixoforming. Therefore, a decrease of the shear rate $\dot{\gamma}$ will not lead to an increase in viscosity η. A sudden increase of the shear rate *will* lead to a decrease in viscosity. In effect, an increase in $\dot{\gamma}$ beyond the largest shear rate $\dot{\gamma}_o$ experienced so far will lead to a decrease in viscosity (and modify the maximum shear rate $\dot{\gamma}_o$). It is this "ratchet-type" behavior, which Orgéas et al. model but they assume a single value of $\dot{\gamma}_o$ because they have a simple geometry involving a transition between a shot sleeve and a small injection aperture. Most of the change in viscosity occurs at that point.

Orgéas et al. obtain the parameters for their model from measuring the pressures and the temperatures in a tube with a shaft in it. The simulation then involves the filling of a reservoir. Figure 7.16 shows comparisons between simulations and interrupted filling experiments for A356 alloy.

Orgéas et al. [93] found eutectic-rich concentric rings in the tube in a Poiseuille type experiment (Fig. 7.17). These were due to veins of liquid formed in the shot sleeve as a result of mechanical instabilities in the solid skeleton, which is not

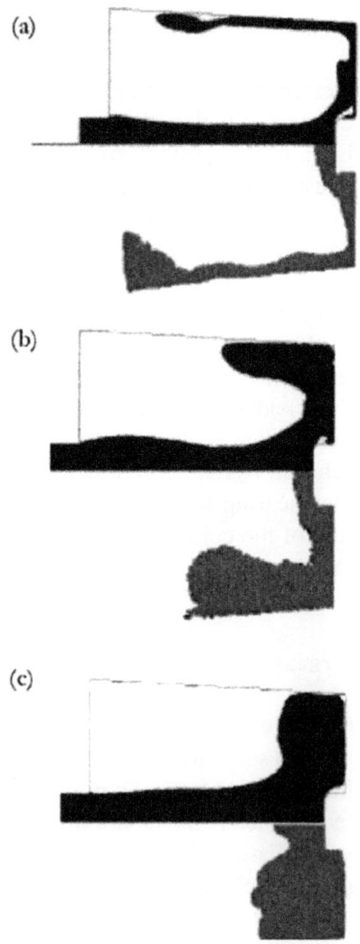

Fig. 7.16 Comparison between experimental and simulated filling of a cavity (initial ram velocity 0.8 ms^{-1}, diameter of tube 25 mm) [93]. (**a**) Fraction solid 0.52. (**b**) Fraction solid 0.58. (**c**) Fraction solid 0.73. In each figure, the *upper part* is the simulation and the *lower part* the experimental result obtained with interrupted filling. The material is flowing into the cavity from the left

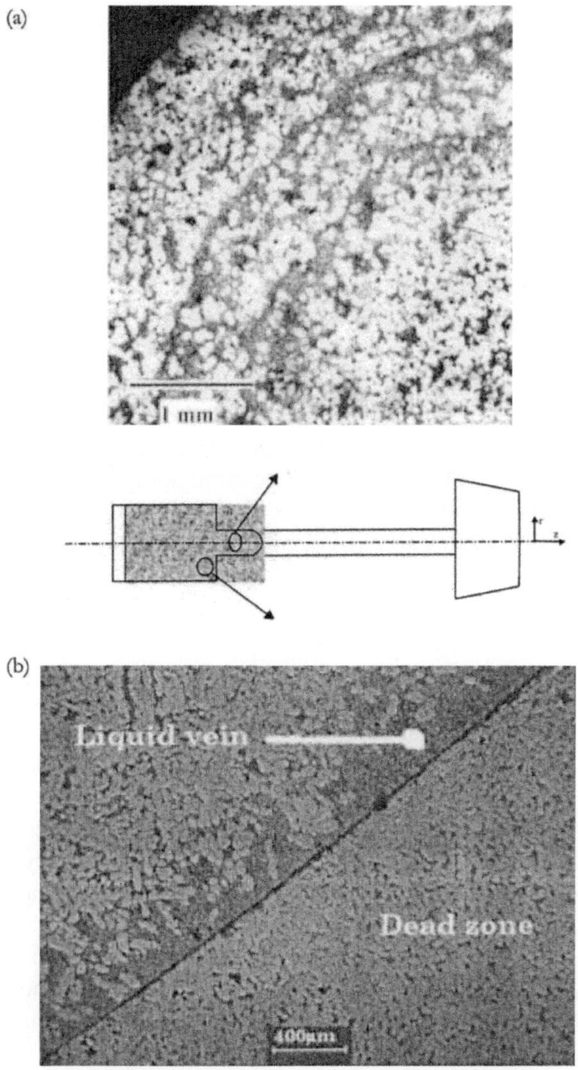

Fig. 7.17 (**a**) Cross-section through the semisolid material solidified in a tube in a Poiseuille-type experiment. The eutectic-rich concentric rings are due to veins of liquid formed in the shot sleeve. (**b**) shows such a vein formed at the limit of the dead zone in the shot sleeve (*see inset*). [93]

uniformly sheared. A "dead zone" formed at the bottom right corner of the shot sleeve, i.e., a zone which hardly suffers any shear and only experiences compression. The compression of the dead zone produces an effect like squeezing a sponge. This type of behavior cannot be predicted with a one-phase model and emphasizes the need for two-phase models, which will be described later. Dead zones need to be avoided in die design for semisolid processing.

Jahajeeah et al. [94] have also used the PLCO model. There is reasonable agreement with the results of interrupted filling tests with A356 aluminum alloy. Defects are predicted with certain runner arrangements and these are found in practice.

7.6.2 Two-phase Finite Element Modeling

Gebelin et al. [95] have presented a useful mathematical comparison of one-phase and two-phase approaches. In addition, there is a detailed introduction by Petera in [96]. Orgéas et al. [93] have also reviewed two-phase approaches.

In the two-phase models, the semisolid material is treated as a saturated two-phase medium, i.e., made of liquid and solid phases. Each phase has its own behavior, which can be influenced by the presence of the other phase via interfacial contributions. The conservation equations can be written within a mixture theory background [97]. The solid phase (which can be thought of as a solid skeleton) can be modeled as a purely viscous and compressive medium [38, 98]. A Darcy-type term appears in the momentum equations to handle the momentum exchanges between the solid and the liquid, which is assumed to be Newtonian [99]. Phase separation (see Fig. 7.11) can be predicted with these models. The principal difficulty, though, is that the determination of the required rheological parameters is not straightforward [38,98]. The computation time also tends to be high because the models require the simultaneous calculation of a solid fraction field, a pressure field, two velocity fields (for the liquid and the solid), and a temperature field (unless the treatment is isothermal).

Zavaliangos [100] uses an internal variable c to represent the degree of cohesion, which evolves with deformation. This is directly analogous with the structural parameter λ in the model of Brown et al. discussed earlier (Sect. 7.3). The behavior is not symmetric under tension and compression (Fig. 7.18). The permeability equation implies that solid–liquid segregation decreases with a reduction in the globule size.

The simulations show that high strain rates result in near undrained conditions and minimal phase segregation. This finding very much reflects the experimental findings, although there are limits. If the strain rate is too high, flow will become turbulent and material will tend to jet across the cavity.

The solid phase is assumed to be a pseudofluid with a Herschel–Bulkley viscosity relationship [101]. The simulation is nonisothermal and has been validated for Sn-15%Pb (Figs. 7.19 and 7.20) with real time visualization of the flow in the die through a transparent window.

Kang and Jung [102] treated the solid phase as compressible and introduced a separation coefficient expressing the actual separation of the particles in relation to their initial separation. As with Zavaliangos [100], the higher the strain rate, the more homogeneous the distribution of the solid fraction. In Binet and Pineau [103], a mixture approach is adopted where the hydrodynamic part is the same as for most incompressible CFD codes, but the velocity field represents the velocities of the mixture. A source term is added to the momentum equations to take account of the diffusion velocities of the different phases. Choi et al. [104] assume the solid to be viscoplastic and use Kuhn's yield criterion for the solid phase (i.e., behavior is symmetric for tension and compression and the hydrostatic component of stress is included). Yoon et al. [105], in contrast, use von Mises yield criterion

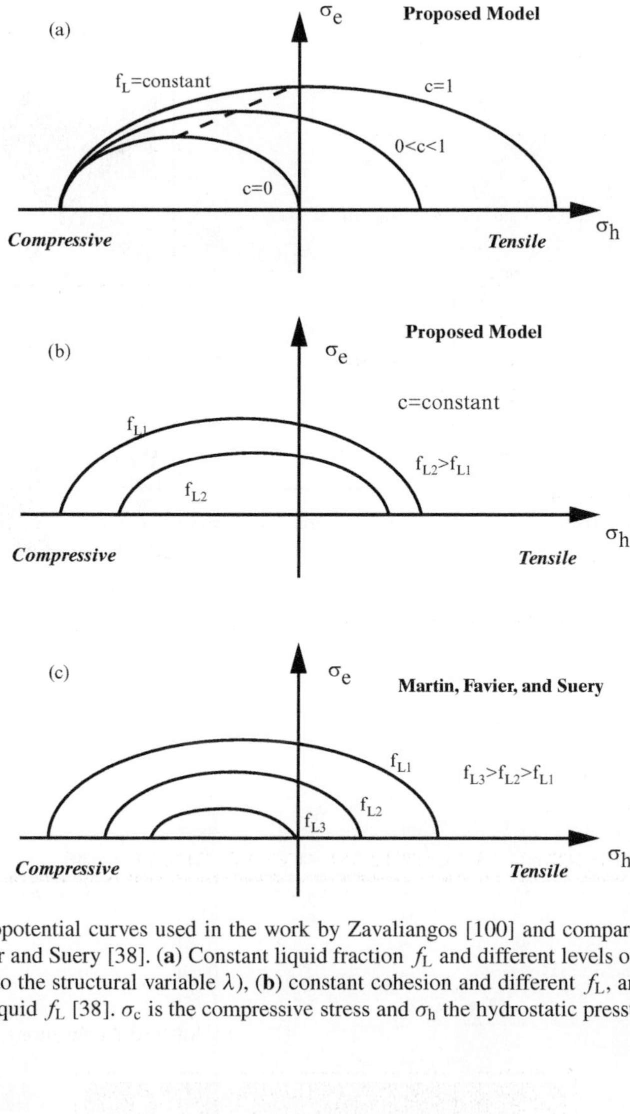

Fig. 7.18 Isopotential curves used in the work by Zavaliangos [100] and compared with that in Martin, Favier and Suery [38]. (**a**) Constant liquid fraction f_L and different levels of cohesion c (c is analogous to the structural variable λ), (**b**) constant cohesion and different f_L, and (**c**) different fractions of liquid f_L [38]. σ_c is the compressive stress and σ_h the hydrostatic pressure in the solid phase

(i.e., symmetric in tension and compression) and treat the semisolid as a single incompressible phase. Kopp and Horst [106] adopt the Drucker–Prager yield criterion (which is asymmetric for tension and compression). Modigell et al. [107] use the pseudofluid approach for the solid phase [108]. All the non-Newtonian properties are shifted to the solid phase and the liquid is Newtonian. Two-dimensional contour maps showing the transitions between laminar, transient, and full turbulent filling are plotted (Fig. 7.21). The three-dimensional process window for A356 aluminum alloy, based on laminar filling is also delineated (Fig. 7.22), providing valuable guidance for actual processing.

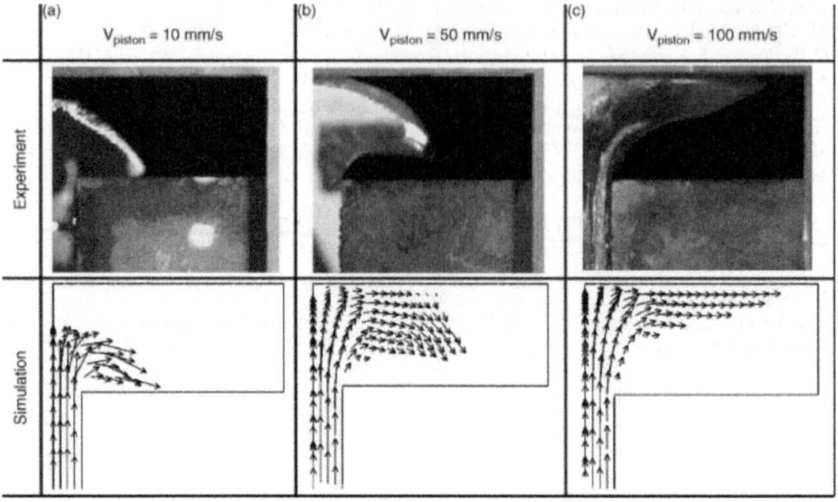

Fig. 7.19 Comparison between the simulated and the observed contours of the flow front for different filling velocities. The simulations have been carried out using one set of model parameters [101]

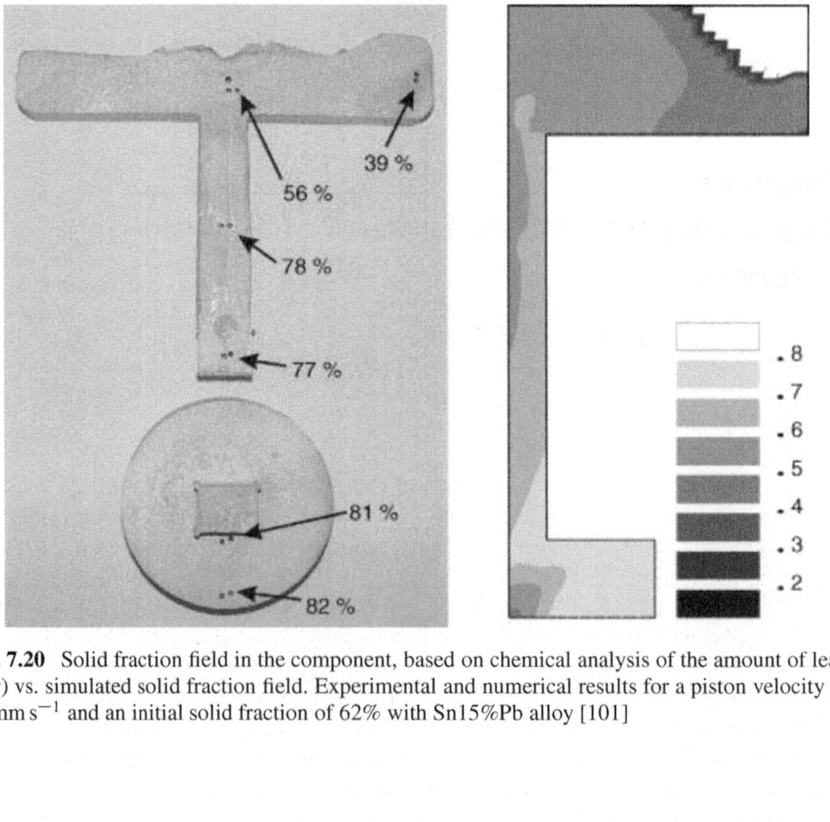

Fig. 7.20 Solid fraction field in the component, based on chemical analysis of the amount of lead (*left*) vs. simulated solid fraction field. Experimental and numerical results for a piston velocity of 50 mm s^{-1} and an initial solid fraction of 62% with Sn15%Pb alloy [101]

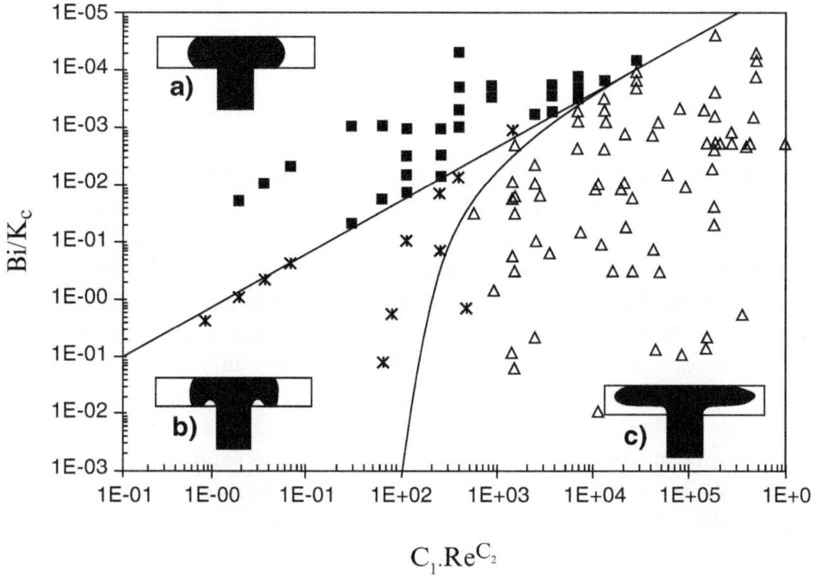

Fig. 7.21 Map of types of flow [107]. (**a**) Laminar, (**b**) Transient, (**c**) Turbulent. *Bi* is the Bingham number, K_c a rheological number, C_1, C_2 geometric constant and *Re* the Reynolds number. K_c, C_1 and C_2 are not specified in the paper

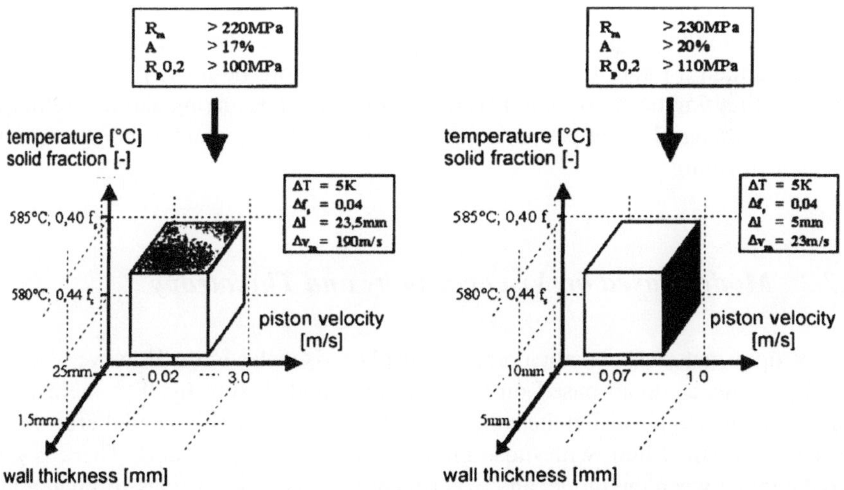

Fig. 7.22 Three-dimensional process window for aluminum alloy A356 [107]. Mechanical properties are given in the *two top boxes*. The *smaller boxes* summarize the process parameter windows to obtain those mechanical properties (ΔT is the temperature window, Δf_s the solid fraction window, Δl the wall thickness and Δv_m). The higher the required mechanical properties, the smaller the three-dimensional process window (compare the *right hand diagram* with the *left*)

7.7 Models akin to Forging Models

7.7.1 Viscoplastic Constitutive Models

Ding et al. [109] used DEFORM-3D software to simulate a rigid viscoplastic alloy being compressed within a six-fingered die heated to the initial temperature of the billet, i.e., the conditions are quasi-isothermal. The researchers fitted the flow stress in the steady state when obtaining the parameters and did not attempt to fit the flow stress during the initial breakdown stage. The die fingers are of different cross-sectional areas. Metal in the larger orifice fingers flows faster than that in the smaller ones. This is contrary to what would be expected with a thixotropic fluid. The simulation is therefore similar to a soft forging simulation. It does, however, include some time dependence in the constitutive equation through a $\dot{\gamma}$ term.

There has been extensive modeling of forging in the solid state and codes, such as FORGE 3 are relatively well established. Some workers (e.g., Rassili et al. [110]) have used viscoplastic constitutive models within such codes to model semisolid forming (Fig. 7.23). The viscoplastic equation is:

$$K\left(\bar{\varepsilon}, T\right) = K_{\mathrm{o}} \exp\left(\frac{\beta}{T}\right) \bar{\varepsilon}^{n}. \tag{7.16}$$

There is no time dependence for the viscosity within these models and these are essentially forging simulations, albeit soft forging. For semisolid processing, where the fraction solid is high (perhaps 80% or more), this may be a valid approach. The process is then a rather different one from where more liquid is present, because the shear thinning behavior of a thixotropic fluid is not occurring and it is rather the softness of material at temperatures just above the solidus, which is being exploited for shape forming.

7.7.2 Model Based on Viscoelasticity and Thixotropy

Thixotropic materials do not normally display viscoelasticity. However, Wahlen [111] presents a model based on viscoelasticity and thixotropy. The inclusion of viscoelasticity suggests that the solid fraction must be rather high and therefore, this model is classified here with those models akin to forging models. There is good agreement between the model and experiment for flow stress versus strain at a temperature of 570°C for A356 aluminum alloy. However, this temperature is such that the alloy would be expected to be nearly fully solid. There is a transition in behavior around 570°C, which the author interprets as the transition between plastic deformation of a connected-particle network and the viscous flow of a suspension of solid particles.

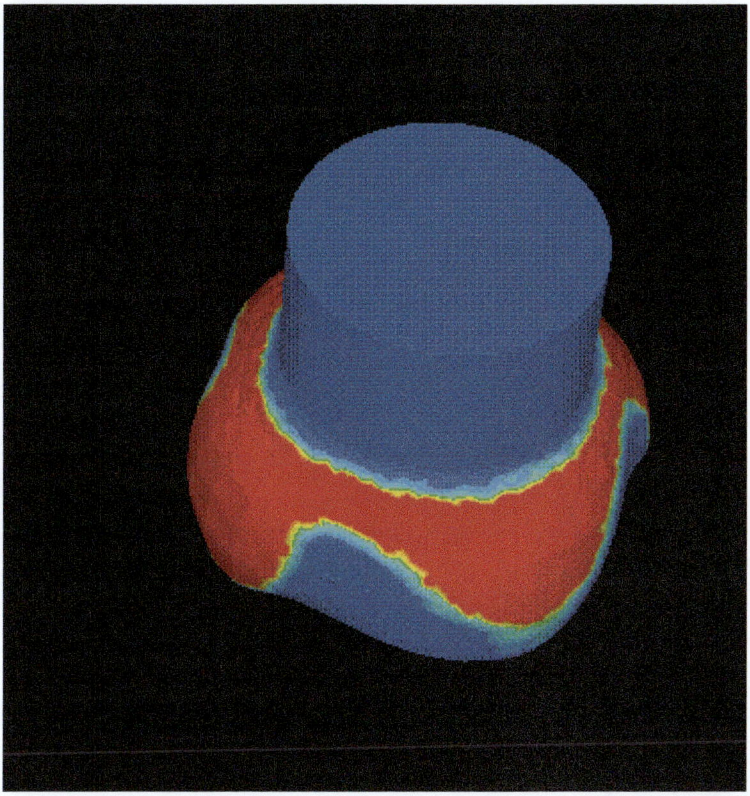

Fig. 7.23 Finite element simulation with FORGE 3 of steel in the semisolid state. The *colors* represent the extent of deformation with *red* high and *blue* low. (Courtesy A. Rassili and coworkers [110])

7.8 Micromodeling

Micromodeling is a novel and interesting approach, which draws on micromechanics and homogenization technique for disordered materials [112]. It is summarized in detail by Favier [113]. It is sometimes referred to as "micro–macro" modeling.

Spherical inclusions (i.e., particles or globules) containing entrapped liquid are assumed to deform very little. They can slip relative to each other if the restriction between them is released. Generally, deformation takes place in the "active zone" around the inclusions, which consists of the solid bonds and that liquid, which is not entrapped. The microstructure is therefore represented as shown in Fig. 7.24, with, in effect, all the inclusions gathered into a single inclusion and this single inclusion coated with the "active zone." The coated inclusion is then embedded into a matrix having the effective properties of the heterogeneous semisolid material (the "homogeneous equivalent medium").

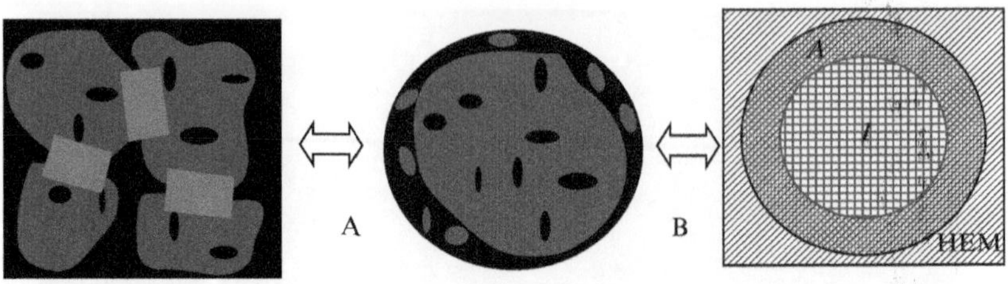

Fig. 7.24 (**a**) shows a schematic representation of the semisolid microstructure showing the presence of one inclusion of solid and entrapped liquid surrounded by a coating of liquid and solid bonds. In (**b**), a coated inclusion is embedded into a matrix having the effective properties of the heterogeneous semisolid material. This is called the "homogeneous equivalent medium" (HEM). [113]

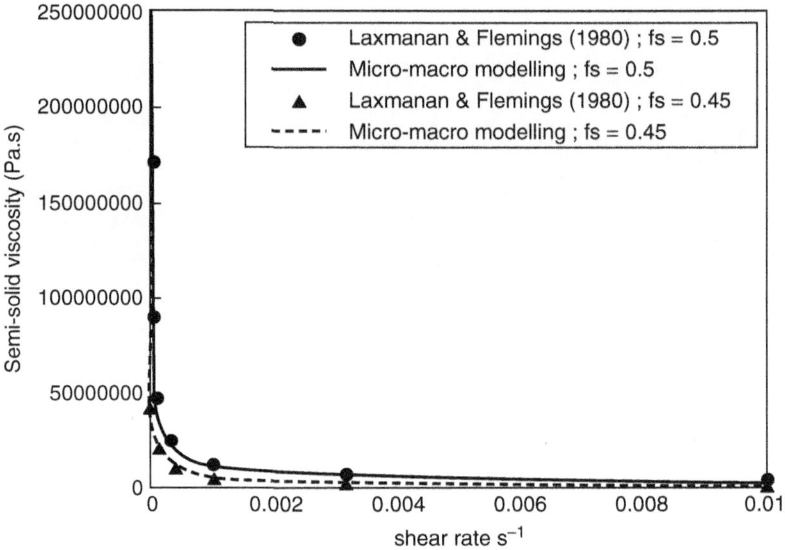

Fig. 7.25 Comparison of "micro–macro" modeling with experimental results for "low" shear rates with Sn15%Pb alloy compressed between parallel plates [45, 113]

The volume solid fraction of the active zone is the internal variable. During deformation, bonds are broken and liquid is released. The liquid–solid distribution therefore changes with strain rate. The liquid and the solid are both assumed isotropic and incompressible. Predictions are compared with experiment in Figs. 7.25 and 7.26. Figure 7.25 is for "low" shear rates and Fig. 7.26 for "high" shear rates, although these are still relatively low in comparison with shear rates in the die in real processing. Note that the parameters for the two fractions of solid in Fig. 7.26 are the same. This is a key test of modeling where often "fitting" of experimental results under one set of conditions is used to obtain the parameters for modeling. To be useful, a model must then be able to accurately predict behavior under a different

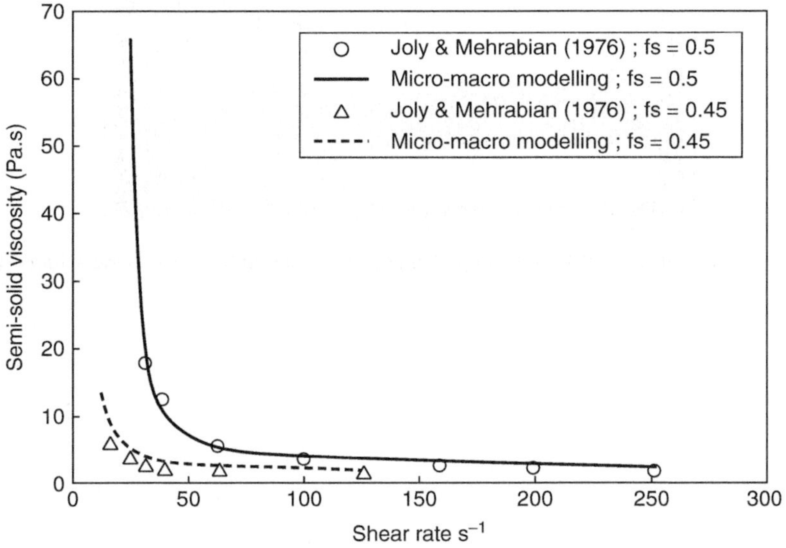

Fig. 7.26 Comparison between "micro–macro" modeling and experimental results for "high" shear rates for Sn15%Pb alloy in concentric cylinder viscometer tests [2, 113]

set of conditions without a "refitting" procedure being necessary. This model is demonstrating this capability.

7.9 Concluding Remarks on Modeling

Modeling of semisolid processing is challenging. It sits on the interface between solid deformation modeling and fluid mechanics, hence placing demands both in mathematical terms, in terms of obtaining the input material parameters and also in terms of sheer computing power. Here, the utility of modeling has been illustrated and issues of whether to use a yield stress in modeling have been highlighted. The use of a structural variable has been described. Models have then been categorized as finite difference or finite element and as one-phase or two-phase. In addition, models which treat semisolid processing as akin to soft forging have been identified. Finally, the approach called "micromodeling" has been outlined.

All modeling must be constantly challenged and validated by experimental results. This can be a matter of comparing with viscometry results (which tend to be for relatively low shear rates) or with compression experiments between parallel plates (where higher shear rates can be achieved). Alternatively, comparison can be made with real die filling. Many workers have carried out interrupted filling tests but inertia tends to continue to carry the material forwards after the stroke has stopped and therefore, the results of interrupted filling must be treated with circumspection. Some real time flow visualization is now available. Where experimental validation is available, this has been identified in this Chapter.

Chapter 8
General Conclusions on Rheology and Modeling

Although semisolid forming is far from replacing conventional casting and forging processes, it has become interesting for the production of specific parts for various industrial sectors. In order to avoid time-consuming trial-and-error experiments to design the mould or the die and to determine the optimum forming conditions, numerical simulation is now extensively developed. It requires an important research activity both to improve the simulation methods for better accuracy and reduced computer time and to develop models that reproduce at best the behavior of the material during the forming operation. In Part II, the experiments that have been developed to determine the rheological behavior of alloys in the semisolid state have been described together with the representative results of these experiments. The models which have been proposed to account for this behavior have also been detailed. The behavior of this type of material is now much better understood but there are still debates concerning particular aspects, such as the existence of a yield stress. In addition, many researchers are still studying the behavior of the alloy under steady state conditions to derive models to be applied for highly transient situations. Finally, under most forming conditions, the alloy behaves like a suspension of interacting solid particles so that this type of model can reasonably predict the filling of the mould and the optimum forming parameters. Development of two-phase models has been carried out more recently, but there is still much work to be done to introduce them in computer codes for the prediction of segregation during a forming operation. However, whatever be the type of model, pertinent microstructural parameters have to be brought in, which still requires some effort for their determination. On this aspect, X-ray microtomography is a powerful tool, which should be particularly appropriate for this determination despite the relative demands of the experiments.

David H. Kirkwood et al., *Semi-solid Processing of Alloys*,
Springer Series in Materials Science,
DOI 10.1007/978-3-642-00706-4_8,

References for Part II Rheology and Modeling

1. D.B. Spencer, R. Mehrabian, M.C. Flemings, Metall. Trans. **3**, 1925 (1972)
2. P.A. Joly, R. Mehrabian, J. Mater. Sci. **11**, 1393 (1976)
3. C.J. Quaak, Ph.D. thesis, TU Delft, 1996
4. S.A. Metz, M.C. Flemings, Trans. Am. Foundrymen's Soc. **77**, 329 (1969)
5. S.A. Metz, M.C. Flemings, Trans. Am. Foundrymen's Soc. **77**, 453 (1969)
6. S.M. Nabulsi, T.A. Steinberg, C.J. Davidson, N.W. Page, in *Proceedings of 4th International Conference On Semi-Solid Processing of Alloys and Composites*, UK, 1996
7. T. Sumitomo, D.H. StJohn, T. Steinberg, Mater. Sci. Eng. **A289**, 18 (2000)
8. O. Ludwig, J.-M. Drezet, C.L. Martin, M. Suéry, Met. Mater. Trans. **36A**, 1525 (2005)
9. O. Lashkari, S. Nafisi, R. Ghomashchi, Mater. Sci. Eng. **A441**, 49 (2006)
10. O. Lashkari, R. Ghomashchi, Mater. Sci. Eng. **A454**, 30 (2007)
11. W.R. Loué, Evolution microstructurale et comportement rhéologique d'alliages Al-Si à l'état semi-solide, Thèse INP Grenoble, 1992
12. T.Z. Kattamis, T.J. Piccone, Mater. Sci. Eng. **A131**, 265 (1991)
13. H.-I Chen, J.-C. Chen, J.-J. Liao, Mater. Sci. Eng. **A487**, 114 (2008)
14. L.S. Turng, K.K. Wang, J. Mater. Sci. **26**, 2173 (1991)
15. Y. Ito, M.C. Flemings, J.A. Cornie, in *Nature and Properties of Semi-Solid Materials*, ed. by J.A. Sekhar, J. Dantzig, (The Minerals, Metals and Materials Society, Warrendale, PA, 1991), p. 3
16. S. Verrier, M. Braccini, C. Josserond, L. Salvo, M. Suéry, W. Ludwig, P. Cloetens, J. Baruchel, in *X-Ray Tomography in Materials Science*, ed. by J. Baruchel, J.-Y. Buffière, E. Maire, P. Merle, G. Peix, (Hermes Science, Paris, 2000), p. 77
17. A.R.A. McLelland, N.G. Henderson, H.V. Atkinson, D.H. Kirkwood, Mater. Sci. Eng. **A232**, 110 (1997)
18. J. Koke, M. Modigell, J. Non-Newtonian Fluid Mech. **112**, 141 (2003)
19. D.H. Kirkwood, P.J. Ward, Mater. Lett. **62**, 3981 (2008)
20. W.R. Loué, S. Landkroon, W.H. Kool, Mater. Sci. Eng. **A151**, 255 (1992)
21. P. Kumar, C.L. Martin, S. Brown, Metall. Trans. **24A**, 1107 (1993)
22. B.P. Gautham, P.C. Kapur, Mater. Sci. Eng. **A393**, 223 (2005)
23. T.Y. Liu, H.V. Atkinson, P.J. Ward, D.H. Kirkwood, Met. Mater. Trans. **34A**, 409 (2003)
24. H.K. Moon, Ph.D. thesis, MIT, 1990
25. A. Assar, N. El-Mahallawy, M.A. Taha, Aluminium **57**, 807 (1981)
26. C.J. Paradies, M. Rappaz, T. Imwinkelried, J.P. Gabathuler, in *Proceedings of 4th International Conference On Semi-Solid Processing of Alloys and Composites*, UK, 1996
27. D. Bernhard, H. Müller-Späth, P.R. Sahm, in *Proceedings of 5th International Conference On Semi-Solid Processing of Alloys and Composites*, Golden, USA, 1998
28. S. Sannes, H. Gjestland, L. Arnberg, J.K. Solberg, in *Proceedings of 3rd International Conference on Semi-Solid Processing of Alloys and Composites*, Japan, 1994
29. M. Mada, F. Ajersch, in *Proceedings of 2nd International Conference On Semi-Solid Processing of Alloys and Composites*, Cambridge, USA, 1992

30. J.A. Yurko, M.C. Flemings, in *Proceedings. Of the Merton C. Flemings Symposium on Solidification and Materials Processing*, ed. by R. Abbaschian, H. Brody et A. Mortensen, (TMS, Warrendale USA, 2001), p. 281
31. J.A. Yurko, PhD. Thesis, Massachusetts Institute of Technology, 2001
32. J.A. Yurko, M.C. Flemings, Met. Mat. Trans. **33A**, 2737 (2002)
33. E.A. Vieira, A.M. Kliauga, M. Ferrante, J. Mat. Proc. Technol. (2004) 1623
34. E.A. Vieira, M. Ferrante, Acta Mater. **53**, 5379 (2005)
35. P. Vicente-Hernandez, F. Decultieux, P. Schmidt, I.L. Svensson, C. Levaillant, ISIJ Int. **35**, 805 (1995)
36. R. Bigot, V. Favier, C. Rouff, J. Mat. Proc. Technol. **160**, 43 (2005)
37. T.G. Nguyen, M. Suéry, D. Favier, *Proceedings of 2nd International Conference. On Semi-Solid Processing of Alloys and Composites*, Cambridge, USA, 1992
38. C.L. Martin, D. Favier, M. Suéry, Int. J. Plasticity **13**, 237 (1997)
39. J.C. Gébelin, M. Suéry, D. Favier, Mater. Sci. Eng. **A272**, 134 (1999)
40. G.-C. Gullo, K. Steinhoff, P.J. Uggowitzer, in *Proceedings of 6th International Conference on Semi-solid Processing of Alloys and Composites*, ed. by G.L Chiarmetta, M. Rosso (Edimet, Brescia, 2000) p. 367
41. L. Nguyen Thanh, M. Suéry, Mater. Sci. Technol. **10**, 894 (1994)
42. L. Salvo, W.R. Loué, M. Suéry, ISIJ Int. **35**, 798 (1995)
43. M. Suéry, M.C. Flemings, Metal. Trans. **13A**, 1809 (1982)
44. J. Valer, P. Ménéses, F. Saint-Antonin, M. Suéry, Mater. Sci. Eng. **A272**, 342 (1999)
45. V. Laxmanan, M.C. Flemings, Metall. Trans. **11A**, 1927 (1980)
46. T.G. Nguyen, Thèse INP Grenoble, 1991
47. O. Ludwig, M. DiMichiel, P. Falus, L. Salvo, M. Suéry, in *Proceedings of 8th International Conference on Semi-solid Processing of Alloys and Composites*, Limassol, Cyprus, 2004
48. J.C. Gébelin, Thèse INP Grenoble, 2000
49. D. Fabregue, A. Deschamps, M. Suéry, W.J. Poole, Met. Mater. Trans. **37**, 1459 (2006)
50. N. Limodin, L. Salvo, M. Suéry, M. Dimichiel, Acta Mater. **55**, 3177 (2007)
51. M. Modigell, L. Koke, J. Mater. Proc. Technol. **111**, 53 (2001)
52. D.H. Kirkwood, P.J. Ward, Steel Res. Int. **75**, 519 (2004)
53. K.J. Burke, PhD Thesis, University of Sheffield, 1998
54. J.H. Ferziger, M. Peric, *Computational Methods for Fluid Dynamics*, 2nd edn. (Springer-Verlag, Berlin Heidelberg New York, 1997)
55. H.K. Versteeg, W. Malalasekera, *An Introduction to Computational Fluid Dynamics* (Longman, New York, 1995)
56. T.Y. Liu, H.V. Atkinson, P. Kapranos, D.H. Kirkwood, S.C. Hogg, Metall Mater Trans A. **34A**, 1545 (2003)
57. A.N. Alexandrou, in *Science and Technology of Semi-Solid Processing*, ed. by A. De Figueredo (Publ. North American Die Casting Association, Rosemont, IL, 2001)
58. Z. Fan, Int. Mater. Rev. **47**, 49 (2002)
59. H.V. Atkinson, Prog. Mater. Sci. **50**, 341 (2005)
60. H.V. Atkinson (ed.), *Modelling of Semi-Solid Processing*, (Aachen, Germany, 2008)
61. H.V. Atkinson, in *Proceedings of International Conference on Synthesis of Lightweight Metals III*, ed by F.H. Froes, C.M. Ward-Close, P.G. McCormick, D. Eliezer (Published by Minerals, Metals and Materials Society, PA, USA, 1999), p. 205
62. H.A. Barnes, J. Non-Newtonian Fluid Mech. **70**, 1 (1997)
63. F. Moore, Trans Br. Ceram. Soc. **58**, 470 (1959)
64. D.C.-H. Cheng, F. Evans, Br. J. Appl. Phys. **16**, 1599 (1965)
65. P. Kumar, C.L. Martin, S. Brown, Acta Metall. Mater. **42**, 3595 (1994)
66. C.L. Martin, P. Kumar, S. Brown, Acta Metall. Mater. **42**, 3603 (1994)
67. O.J. Ilegbusi, S. Brown, J. Mater. Eng. Perform. **4**, 486 (1995)
68. O.J. Ilegbusi, K.A. Quach, M.D. Mat, J. Mater. Eng. Perform. **8**, 31 (1999)
69. A. Zavaliangos, A. Lawley, J. Mater. Eng. Perform. **4**, 40 (1995)
70. G.P. Backer, JOM **50**, 8 (1998)

71. M.R. Barkhudarov, C.L. Bronisz, C.W. Hirt, in *Proceedings of 4th International Conference on Semi-Solid Processing of Alloys and Composites*, ed. by D.H. Kirkwood, P. Kapranos, (University of Sheffield, Sheffield, U.K, 1996), p. 110
72. M.R. Barkhudarov, C.W. Hirt, in *Proceedings of 'Materials Week 96' TMS Conference*, Cincinatti, Ohio, 1996
73. M. Modigell, J. Koke, Mech. Time Dependent Mater. **3**, 15 (1999)
74. J. Crank, *Mathematics of Diffusion*, (Oxford University Press, Oxford, England, 1956)
75. P.J. Ward, H.V. Atkinson, D.H. Kirkwood, S.B. Chin, T.Y. Liu, in *Proceedings of International Conference Modelling of Casting, Welding and Advanced Solidification Processes IX (MCWASP IX)*, Aachen, Germany, 2000
76. P.J. Ward, H.V. Atkinson, D.H. Kirkwood, S.B. Chin, Final Report for Engineering and Physical Sciences Research Council Project "Modelling Thixotropic Flow of Metal Alloys into a Die" (GR/M17334/01) http://www.le.ac.uk/engineering/staff/atkinson_files/modelling_final_report.pdf, 2002
77. G. Messmer, in *Proceedings of 7th International Conference on Advanced Semi-Solid Processing of Alloys and Composites*, Tsukuba, Japan, ed. by Y. Tsutsui, M. Kiuchi, K. Ichikawa, (National Institute of Advanced Industrial Science and Technology and the Japan Society for Technology of Plasticity, Japan, 2002), p. 527
78. A. Palekar, A. Starobin, A. Reikher, in *Proceedings of 68th World Foundry Congress*, Chennai, India, 2008
79. N.S. Kim, C.G. Kang, J. Mater. Process. Technol. **103**, 237 (2000)
80. P.K. Seo, C.G. Kang, in *Proceedings of 7th International Conference on Advanced Semi-Solid Processing of Alloys and Composites*, ed. by Y. Tsutsui, M Kiuchi, K. Ichikawa, (National Institute of Advanced Industrial Science and Technology and the Japan Society for Technology of Plasticity, Tsukuba, Japan, 2002), p. 515
81. M. Itamura, M. Adachi, K. Murakami, T. Harada, M. Tanaka, S. Sato, T. Maeda, Int. J. Cast Met. Res. **15**, 173 (2002)
82. A. Alexandrou, F. Bardinet, W. Loué, J. Mater. Process. Technol. **96**, 59 (1999)
83. G.R. Burgos, A.N. Alexandrou, J. Rheol. **43**, 485 (1999)
84. Q.Y. Pan, D. Apelian, A.N. Alexandrou, Metall. Mater. Trans. B **35B**, 1187 (2004)
85. G.R. Burgos, A.N. Alexandrou, V. Entov, J. Mater. Process. Technol. **110**, 164 (2001)
86. A.N. Alexandrou, Y. Pan, D. Apelian, G. Georgiou, in *Proceedings of 7th International Conference on Advanced Semi-Solid Processing of Alloys and Composites*, ed. by Y. Tsutsui, M. Kiuchi, K. Ichikawa, (National Institute of Advanced Industrial Science and Technology and the Japan Society for Technology of Plasticity, Tsukuba, Japan, 2002), p. 417
87. G.C. Florides, A.N. Alexandrou, G.C. Georgiou, J. Non-Newtonian Fluid Mech. **143**, 38 (2007)38
88. A.N. Alexandrou, P. Le Menn, G. Georgiou, V. Entov, J. Non-Newtonian Fluid Mech. **116**, 19 (2003)
89. T.C. Papanastasiou, J. Rheol. **31**, 385 (1987)
90. B. Niroumand, K. Xia, Mater. Sci. Eng. A **283A**, 70 (2000)
91. E. Tzimas, A. Zavaliangos, J. Mater. Sci. **35**, 5319 (2000) 5319
92. A.N. Alexandrou, G. Georgiou, J. Non-Newtonian Fluid Mech. **142**, 199 (2007)
93. L. Orgéas, J-P. Gabathuler, Th. Imwinkelried, Ch. Paradies, M. Rappaz, Modell. Simul. Mater. Sci. Eng. **11**, 553 (2003)
94. N. Jahajeeah, R. Bruwer, O. Damm, L. Ivanchev, P. Rossouw, K. Sharma, in *Proceedings of 7th International Conference on Advanced Semi-Solid Processing of Alloys and Composites*, ed. by Y. Tsutsui, M. Kiuchi, K. Ichikawa, (National Institute of Advanced Industrial Science and Technology and the Japan Society for Technology of Plasticity, Tsukuba, Japan, 2002), p. 533
95. J.C. Gebelin, D. Favier, M. Suéry, in *Proceedings of 5th International Conference on Semi-Solid Processing of Alloys and Composites*, ed. by A.K. Bhasin, J.J. Moore, K.P. Young, S. Midson, (Colorado School of Mines, Golden, Colorado, USA, 1998), p. 309
96. J. Petera, *Modelling of Semi-Solid Processing*, Chap. 4, ed. by H.V. Atkinson (Aachen, Germany, 2008)

97. R.M. Bowen, *Continuous Physics. III, Mixture and EM Field Theories*, (Academic, New York, USA, 1976)
98. T.Y. Nguyen, D. Favier, M. Suéry, Int. J. Plasticity, **10**, 663 (1994)
99. J.C. Gebelin, C. Geindreau, L. Orgéas, P. Royer, D. Favier, J.L. Auriault, in *Proceedings 6th International Conference on Semi-Solid Processing of Alloys and Composites*, ed. by G.L. Chiarmetta, M. Rosso , (Edimet Spa, Brescia, Italy, 2000), p. 155
100. A. Zavaliangos, Int. J. Mech. Sci. **40**, 1029 (1998)
101. M. Hufschmidt, M. Modigell, J. Petera, J. Non-Newtonian Fluid Mech. **134**, 16 (2006)
102. C.G. Kang, H.K. Jung, Int. J. Mech. Sci. **41**, 1423 (1999)
103. B. Binet, F. Pineau, in *Proceedings of Mathematical Modelling in Metal Processing and Manufacturing-COM 2000, Ottawa, Ontario, Canada*, ed. by P. Martin, S. MacEwen, Y. Verreman, W. Lui, J. Goldak, (Publ. Canadian Metallurgical Society, Canada, 2000)
104. J.C. Choi, J.H. Park, B.M. Kim, J. Mater. Process. Technol. **105**, 49 (2000)
105. J.H. Yoon, Y.T. Im, N.S. Kim, J. Mater. Process. Technol. **113**, 153 (2001)
106. R. Kopp, M. Horst, in *Proceedings of 7th International Conference on Advanced Semi-Solid Processing of Alloys and Composites*, ed. by Y. Tsutsui, M. Kiuchi, K. Ichikawa, (National Institute of Advanced Industrial Science and Technology and the Japan Society for Technology of Plasticity, Tsukuba, Japan, 2002), p. 719
107. M. Modigell, R. Kopp, P.R. Sahm, D. Neuschutz, J. Petera, in *Proceedings of 7th International Conference on Advanced Semi-Solid Processing of Alloys and Composites*, ed. by Y. Tsutsui, M. Kiuchi, K. Ichikawa, (National Institute of Advanced Industrial Science and Technology and the Japan Society for Technology of Plasticity, Tsukuba, Japan, 2002), p. 77
108. J. Koke, M. Modigell, J. Petera, Appl. Mech. Eng. **4** (Special Issue ICER'99), 345 (1999)
109. Z.Y. Ding, H.P. Pan, S.S. Xie, in *Proceedings of 7th International Conference on Advanced Semi-Solid Processing of Alloys and Composites*, ed. by Y. Tsutsui, M. Kiuchi, K. Ichikawa, (National Institute of Advanced Industrial Science and Technology and the Japan Society for Technology of Plasticity, Tsukuba, Japan, 2002), p. 455
110. A. Rassili, G. Geuzaine, P. Dular, M. Robelet, J. Demeurger, D. Fischer, in *Proceedings of 8th ESAFORM International Conference on Material Forming*, vol.II, ed. by D. Banabic, (The Publishing House of the Romanian Academy, Bucharest, 2005), p. 1095
111. A. Wahlen, Mater. Sci. Forum **396–402** 185 (2002)
112. C. Rouff, V. Favier, R. Bigot, M. Berveiller, M. Robelet, in *Proceedings of 7th International Conference on Advanced Semi-Solid Processing of Alloys and Composites*, ed. by Y. Tsutsui, M. Kiuchi, K. Ichikawa, (National Institute of Advanced Industrial Science and Technology and the Japan Society for Technology of Plasticity, Tsukuba, Japan, 2002), p. 423
113. V. Favier, *Modelling of Semi-Solid Processing*, Chap. 6, ed. by H.V. Atkinson (Aachen, 2008)

Part III
Industrial Applications of Semisolid Processing

Ken Young and Plato Kapranos

Chapter 9
Introduction to Industrial Applications of Semisolid Processing

Semisolid metal (SSM) processing is a hybrid technology combining features of both casting and forging that enables the production of near net-shape components of superior properties and surface finish. It was developed, from a discovery made at the Massachusetts Institute of Technology in the early 1970s, by Spencer et al. [1] that stirring of alloys during solidification led to a change in the solidifying microstructure resulting not only in a dramatic lowering of the apparent viscosity of the semisolid slurry, but also facilitating two-phase homogeneous flow at quite high fractions solid. More detail of this discovery and the effects of shear rate cooling rate and fraction solid can be found in Part II.

The process of stirring alloys during solidification to produce non-dendritic solid within a slurry, and then injecting this slurry directly into a die as in liquid metal die casting, was called "rheocasting" by the MIT researchers and that name has largely stuck. Rheocasting started out as the preferred process route for industrial production and a new company formed by MIT and a group of industrial partners "Rheocast Corp." designed and built several large-scale rheocasters for production of both aluminium and copper-base alloys. The first major customer of the technology and for the large-scale rheocasters produced by Rheocast Corp. was International Telephone and Telegraph (ITT) Corp.

Initial commercial production at ITT Corp. focussed on copper alloy electrical connector hardware, but soon after commercial production began, lightweight automotives became a focus of attention due to the energy crises, and intense commercial interest began to turn toward aluminium alloys. Almost immediately, questions regarding the process reliability and consistency of rheocasting structural aluminium alloy parts were raised both regarding the viability of large-scale rheocasters in an aluminium cast-house and because rheocast aluminium parts evidenced internal flaws. This spurred development by the ITT corporation of the second alternative process route proposed by MIT, which comprised producing rheocast billet and allowing it to completely cool before re-heating back into the semisolid temperature range for forming or casting.

This alternative process, whereby the slurry is first cast as a fully solid billet, cut into appropriately sized slugs, and subsequently reheated back to the semisolid condition before injecting into a closed die was termed "thixocasting" by the same MIT team. The thixocasting name has endured better in Europe than in the USA

David H. Kirkwood et al., *Semi-solid Processing of Alloys*,
Springer Series in Materials Science,
DOI 10.1007/978-3-642-00706-4_9,

and a similar process of shaping between closed dies was termed "thixoforging". Over the years, proprietary interests have created a wide range of alternative process nomenclature, such as "soft" or "viscous" forging, but primarily the terms semisolid casting or processing are most commonly used. For the purpose of this chapter, all these variants for shaping alloys are collectively referred to as "semisolid metal forming processes".

Although the original MIT research had pioneered the concept of using semisolid material (either rheocast or thixocast) as feedstock for horizontal die-casting machines, the early commercial operations were based around unique, vertical downward stroke, forging machines in an attempt to minimize the use of the specialized raw material. The very first commercial applications of the technology for both copper alloy and aluminium alloy parts were in fact, direct closed die semisolid forging processes, in which effectively 100% of the charge material was converted to the final part. However, while attractive from a material yield/cost point of view, this approach proved troublesome from a quality perspective creating intense segregation beneath the forging ram and incorporating oxides from the charge material surface within the interior of the part among other things.

As more companies became involved in the technology, once that the MIT protective patents ran out, the conventional wisdom of marrying available horizontal die-casting technology with this new raw material won the day. A typical horizontal semisolid metal casting cell is shown in Fig. 9.1.

The last vertical downward forging operations expired in the early 2000s, when AEMP Corporation, the final licensor/custodian of the original MIT technology, filed for bankruptcy. By that time, even the proponents of that approach had converted their technology to utilize gates and runners much like die-casting although

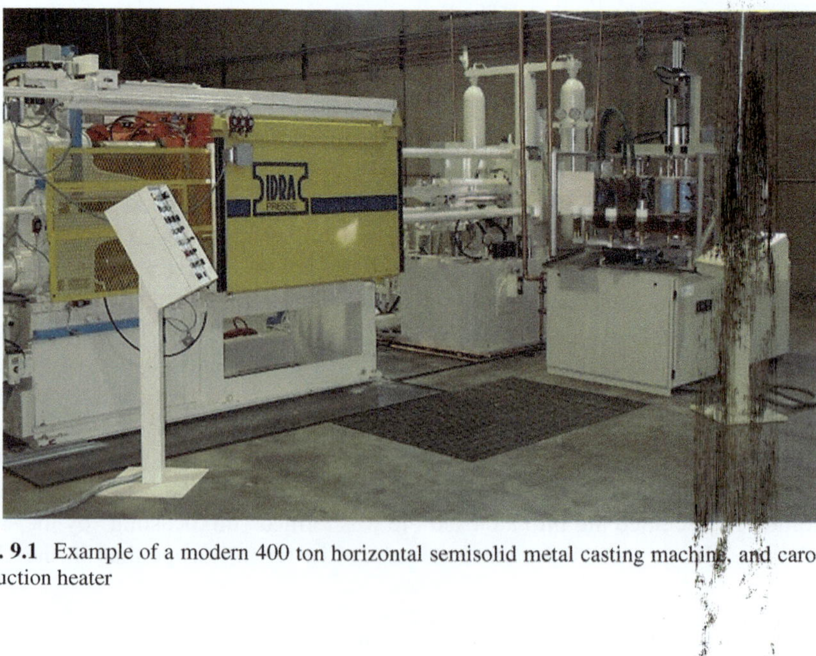

Fig. 9.1 Example of a modern 400 ton horizontal semisolid metal casting machine, and carousel induction heater

in a vertical orientation. Not only was this technology fraught with technical issues like those cited above, but it also suffered from the absence of available infrastructure (who could build tooling for example) and competitive bidding on the very specialized machinery.

In the beginning of commercial operations, both the rheocast and the thixocast process routes were actively pursued, but over a time, the thixocasting process won out for a variety of reasons. For many years, the "thixocasting" process route has dominated the commercial scene. Lately, however, commercial pressure to achieve the lowest possible cost has seen a resurgence of the rheocast approach, since it avoids the need to acquire specially processed billet and allows free purchase of ingot on the open market.

The term "semisolid metal processing" is used here to cover both the operations of producing the raw feedstock and shaping the alloy.

Chapter 10
Raw Material

Non-dendritic slurries were originally produced at MIT by mechanical stirring during solidification of the alloy. This started out as a batch process, but was quickly developed into a continuous slurry generator, or rheocaster, in which liquid alloy was passed through an annular channel while being simultaneously cooled and stirred to exit as a semisolid slurry. In the laboratory, these devices were scaled up to operate all the way up to steel processing temperatures and over the course of some 4 years; MIT researchers cast several thousand steel, aluminium and copper alloy parts using both the rheocasting and thixocasting process routes.

Later, as commercial organisations became more involved, continuous rheocasters capable of producing hundreds of pounds per hour of aluminium and copper alloy slurry were developed and used to cast large quantities of copper alloy parts or aluminium extrusion billet. However, the complexity of operating multiple mechanical stirring units in commercial foundries and concerns about degradation of the equipment and contamination of the product quickly led to the development of electromagnetic (magneto hydrodynamic or MHD) stirring deep in the sump of a DC casting unit. The attraction of the MHD process was the potential to avoid gas pick up while providing vigorous stirring. This has now become the standard industrial route for the manufacture of SSM bar feedstock for use in thixocasting. The original MHD process, developed and patented by Olin Corp. and assigned to ITT Inc., involved rotational flow about the vertical axis of a DC caster (Fig. 10.1).

Although the MHD route provides excellent feedstock for thixoforming, it suffers the cost penalties associated with the need for special processing and a caster who purchases this billet suffers a financial penalty because the "offal" cannot be recycled in house. To circumvent this, renewed interest has been shown in rheocasting, in which slurry production is produced on demand at the die casting unit and several promising commercial processes have been recently developed to meet this need, which will be discussed in detail later (see Sect. 10.6).

David H. Kirkwood et al., *Semi-solid Processing of Alloys*,
Springer Series in Materials Science,
DOI 10.1007/978-3-642-00706-4_10,

Fig. 10.1 Mini horizontal continuous caster equipped with electromagnetic stirrer shown casting 75 mm SSM diameter billet, courtesy Hertwich GmbH

10.1 Semisolid Material Delivery Systems

In production operations around the world, semisolid material is being delivered to the die casting machines either as re-heated billets (slugs) or directly as slurry. For slug re-heating, it is important to note that roughly two-thirds of the re-heating energy is consumed in warming the slug from room temperature to the melting point and only one-third of the energy (although much more time) is consumed during the partial melting stage.

10.2 Vertical Induction Slug Heating

For billet reheating, the dominant technology remains a process, in which vertically oriented slugs are transported sequentially around a carousel of solenoid coils connected in series to an induction power supply. This approach combines a simple mechanical solution with a single, stable, economic and energy conservative power supply, which can be relatively straightforward to regulate (Fig. 10.2).

Each slug steps through each coil and thereby sees an identical heating environment. Heating efficiency can be maximized during heating to the melting range, while lower power coils are used to "soak" the slugs through the semisolid range. Typically, total heating time in minutes, $t_{\min} = \{\text{diameter in mm}/25\}^2$ (or the square of the diameter in inches), in order to permit adequate thermal diffusion from the surface where the induction heat is generated, through the semisolid material to the centre.

Fig. 10.2 Typical carousel induction heater with solenoid coils connected in series to one power supply, courtesy IHS-Inductoheat

Vertical orientation restricts the maximum length to diameter ratio of 2.5:1 and more often 2.25:1 in order for the slugs to remain stable during heating. The slugs must also have a well cut surface, typically saw-cut, in order that they stand upright upon the ceramic pedestals used to support the slugs through the system. However, vertically oriented slugs allow for a simple elimination of the dripping, which occurs during heating of certain alloys. Without care, it is possible for heating to preferentially occur on only one side of the vertical slugs accentuating their tendency to lean over and, in the extreme, to fall off the pedestals.

In order to address this phenomenon, some systems use specially wound coils designed to minimize this differential heating effect. Throughout the heating cycle, the slugs essentially retain their shape, even up to 50% liquid, which facilitates simple pick and place transport of the re-heated slugs to the casting machine. This shape retention also aids removal of the oxide skin, which forms on the slug surface during heating. Because the surface is well-defined, a simple diaphragm device (Fig. 11.1) located at the end of the shot sleeve can serve to strip the oxide from the slug during injection.

10.3 Horizontal Induction Slug Heating

Recently, an alternative horizontal induction heating approach has been promoted mostly within Europe (Fig. 10.3). This is similar in many ways to a convective heating approach, which was perfected for the production of automotive fuel rails several years ago and discussed below. Both horizontal systems utilize boats to contain the slugs that, therefore, no longer require a flat saw-cut surface. At least one facility

Fig. 10.3 Horizontal induction slug heater with individual solenoid coils and lateral slug/boat transport system.

uses a very efficient shearing operation to section billets into slugs prior to heating, thus eliminating all cutting losses.

Unlike the vertical heating technology, each slug is loaded individually into a horizontal solenoid induction-heating coil in which it remains throughout the heating cycle. This approach allows the use of an inductive sensor to detect changes in slug electrical conductivity during heating and thereby compute the degree of melting (electrical conductivity approximately doubles through the melting range for alloy 356). Because the slugs are horizontal, less dripping occurs and the boat retains any drip material, which results from re-heating. This is claimed as an advantage. However, as is also the case of the similar horizontal convective approach, it is difficult to imagine how the dripped-off material is not separated from the slug prior to delivery to the shot sleeve. The consequence is that horizontal heating can result in a less homogeneous charge being delivered to the casting machine.

As in the convective approach, horizontal heating in boats allows higher fraction liquid to be achieved and it is claimed that more complex parts can be produced. However, it is unclear how this technology deals with oxide skins or the inevitable drips or exudation from certain alloys as already mentioned above. Test systems seem to suggest that differential heating associated with the contact area of the slug and boat as well as end effects are also not trivial.

In 2003, a new variation of horizontal induction slug heating was announced, which retains a key advantage of the vertical carousel system in that all slugs see absolutely identical environment as they are heated. This is a horizontal carousel system built around one induction power supply driving a number of horizontal coils connected in series with slugs shuttling into and out of each coil prior to the carousel index. Figure 10.4 shows the outline of one such system designed to heat 125–150 mm diameter slugs up to 550 mm long at a rate of one per minute or approximately 1,600 kg per hour. The power supply is rated at 1,500 kW [2].

Fig. 10.4 Semi-liquid convection slug heating system and cell for fuel rail production

10.4 Horizontal Convection Slug Heating

In the convective oven solution, the mould-washed, steel boats and slugs are transported through a three-zone furnace in several rows (Fig. 10.4). At the exit end of the furnace, a robot is used to pick up the boat, transport it to the shot sleeve and rotate it such that the contents fall directly into the sleeve.

For fuel rails, the fraction liquid can be quite high (typically 60–65%) in order to fill normally quite thin, difficult sections. Since these parts are mostly not heat treated, there is less concern about oxides, which do not generally impact pressure tightness as would porosity.

Although this process extension is not proven in production, it seems very possible to adjust the furnace and delivery system such that higher fraction solid slugs can be developed, which also would facilitate oxide stripping.

Convective heating can be a very energy efficient process and allows a choice of energy sources, which can be important in high electricity cost areas (Fig. 10.5). Convective heating also represents a potential low energy cost pre-heating approach for subsequent induction heating and may be used to raise slug temperatures to a few degrees within the melting range. However, horizontal heating in any form is not trivial and although the boats capture any drip or run-off material, there remains the issue of dealing with this inhomogeneity (Fig. 10.6).

10.5 Reheating of Solid Feedstock

As already discussed, a number of different heating configurations can and have been employed in the reheating of solid feedstock into the semisolid state. For any of these processes, it is critically important that the rate of heat input be controlled so as to maintain a relatively uniform surface to centre temperature profile. Whether

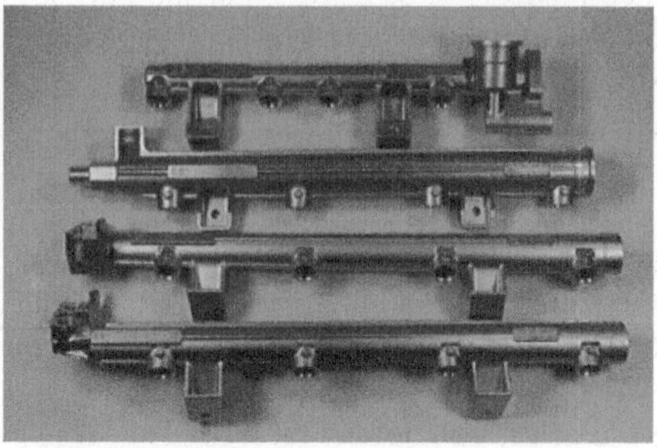

Fig. 10.5 Automotive fuel rails produced using horizontal convective heating and three phase injection die casting machines, courtesy Magnetti- Marelli S.p.A

Fig. 10.6 Horizontal inductively heated slug, alloy 357

the heat source is inductive or convective, the bulk of the heat input arrives at or close to the surface and must be conducted into the centre of the material over time. This process slows substantially once partial melting begins and the relative heat input must be reduced to maintain a reasonable temperature gradient. Too rapid heat input promotes surface melting and accentuates liquid pooling and so-called "run-off", which can be seen in Fig. 10.7a as the "elephant foot" accumulating at the base of the slug.

Mostly, this is not pure surface melting, but rather liquid migration from the surface regions through the porous semisolid material adjacent to the surface. The effect varies from alloy to alloy, with alloys 356/357 typically delivering about 8 wt% run-offs in an adequately heated slug and can be almost zero for alloy 6061 (which has never made serious production due to other issues, such as hot-tearing) and also several copper alloys.

For induction heating, this process is managed by adjusting power levels during heating, with rapid heat input applied, while the material remains fully solid and then reducing the power level after the onset of melting. Fortunately, for alloy such as 356/357, which comprises the bulk of semisolid processing, approximately 65% of the heat input required is consumed within the solid state heating portion with only 35% delivered after the onset of melting (Fig. 10.8).

Fig. 10.7 Illustration of the "elephant foot" effect after the first two-stage heating test (switching time at 140 s) (*left*) and of the semisolid billet keeping its initial shape after the second two-stage heating test (switching time at 205 s) [9]

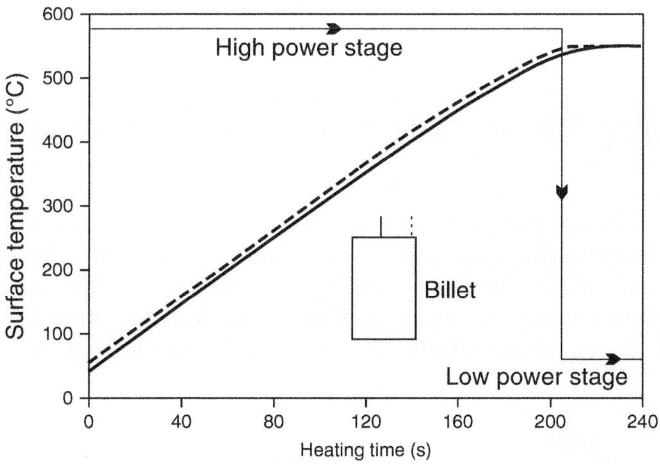

Fig. 10.8 The power input profile for a two-stage heating cycle and the temperature variation on the centre (*large broken lines*) and the edge (*small broken lines*). The heating power was lowered 205 s after heating had started [9]

10.6 Direct Slurry Production: Rheocasting/Thixomoulding

10.6.1 New Rheo-Casting Route

In this process, developed by Ube [3], a fully liquid melt is taken from the holding furnace and poured into specially designed steel cups, which are placed on a carousel next to the new rheo-casting route (NRC) machine. By controlled cooling of the melt into the semisolid region, a globular microstructure is attained without stirring.

Like all variants of direct slurry production, NRC offers the advantage over a billet-based or thixocast route, in that it does not require special precursor material with non-dendritic microstructure. The process starts with normal melt. The melt is prepared in the foundry's melt shop from conventional ingots and then transferred to the holding furnace at the die casting machine. The ladling device pours the melt into the crucibles, which are then put onto the carousel. When the semisolid melt is ready, a robot picks up the crucible and allows the billet to slide into the vertical sleeve of the die casting machine. Theoretically, there is no metal loss in this operation and, after casting, the runners and scrap may be returned to the melt shop for in-house recycling. However, there is invariably a skull of metal left in the container after casting, which must be cleaned out prior to redressing the vessel for the next sequence and cleanly discharging the material from the crucible is not trivial.

While this and other direct slurry approaches allow elimination of the premium precursor material, they do have their own processing issues as discussed. Furthermore, each of the available rheocasting processes also re-introduces liquid metal processing issues back into the semisolid manufacturing cell with the attendant concerns. It is not clear that in the case of NRC whether there is truly any capital cost reduction.

Furthermore, as in any of the direct slurry approaches, each shot becomes a unique solidification event and it is also not clear that all process control issues have been satisfactorily addressed. It is not known at this time whether any significant commercial production of NRC or any other direct slurry manufacturing is occurring. Most likely, for the reasons discussed, the NRC process has not seen widespread adoption despite the fact that it was introduced a decade ago (Fig. 10.9).

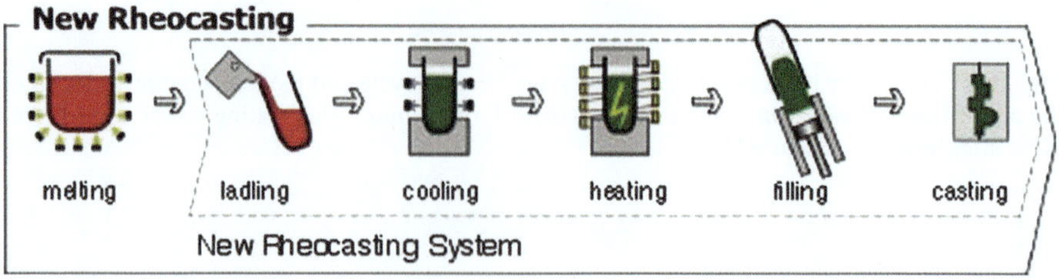

Fig. 10.9 Schematic of the new rheo-casting process (NRC) NRC: pouring – controlled cooling – sleeve filling – forming [3]

10.6.2 Semisolid Rheocasting

Semisolid rheocasting (SSR^{TM}) is another process that forms globular, non-dendritic slurry directly from molten aluminium alloy. The process was developed at the Massachusetts Institute of Technology in 2000 and is now under license to Idra casting machines (ICM) of Brescia, Italy. ICM is a global provider of die casting equipment and sells SSR as a peripheral piece of equipment to its die casting customers.

In the SSR process, molten aluminium alloy is rapidly cooled via a spinning "cold finger" device made from a material, such as graphite, which acts as a heat-sink. As the temperature of the melt is reduced to below the liquidus temperature, nucleation occurs and particles begin to form. Cooling and agitation ceases after the temperature is just below the liquidus and much like in the NRC process, the nuclei grow to form particles. A schematic representation of this process is shown in Fig. 10.10 [4].

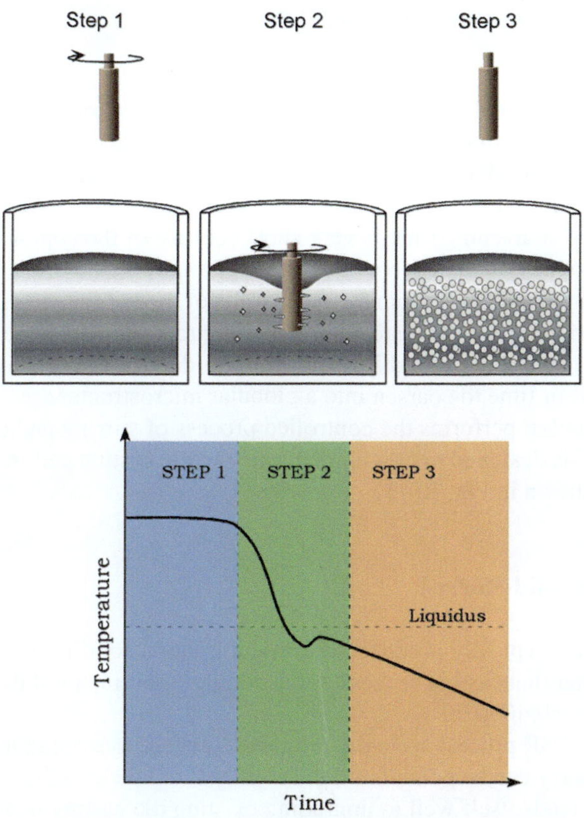

Fig. 10.10 Schematic of the SSR process. Molten alloy is held above the liquidus (Step 1), then rapidly cooled and agitated for a controlled duration to a temperature below the liquidus (Step 2) before agitation and cooling ceases (Step 3).

Fig. 10.11 SSR cell and graphite stirring rod in melt

The rapid cooling and convection of the alloy as it cools through the liquidus temperature promotes formation of very fine, equiaxed dendrites that rapidly coarsen into globular, spheroidal primary phase (See Chapter 4 'Recent Development in Slurry Formation') [5].

The required coarsening time is very short because of the copious, fine dendrite formation. Cooling originates from the moving, rotating cylindrical surface ("cold finger") within the melt and not in the outer walls of the holding vessel. It is claimed that external cooling would create a rapid cooling rate stagnant zone near the walls of the vessel that would be highly susceptible to dendritic formation, and require lengthy periods of time to coarsen into a globular microstructure.

The SSR Station performs the controlled process of stirring and cooling molten aluminium and is designed to fit within an existing die casting cell. A photograph of a SSR cell is shown in Fig. 10.11.

Low-Fraction Solid Slurry

The SSR process typically produces slurry at a fraction solid (<0.20) less than the majority of other rheocasting processes that continue cooling until the fraction solid is approximately 0.40–0.60.

Because the SSR process typically produces low fractions solid, it has the advantage that the slurry can be poured into the normal shot hole of a die casting machine and therefore, lends itself well to upgrading existing die casting operations. Unlike thixo processing, it does not require an extended injection stroke to accommodate the long thixocast billet, but of course, it offers only a reduced advantage in higher viscosity, reduced turbulence and lower solidification shrinkage and heat production upon solidification.

Since the slurries produced are of low fractions solid, alloys that lie closer to eutectic compositions can be employed without formation of the detrimental secondary phases, which start to precipitate at low cooling rates at higher fractions solid. However, many of the alloys that lie close to eutectic compositions, such as A380. 383, 360 are challenged to provide superior mechanical performance because of the iron and other elements they contain and so while SSR processing offers an opportunity to produce sounder, more dense parts, they are typically not necessarily of superior mechanical performance. Because of the reduced heat content of the semisolid alloys, however, the SSR process does offer an opportunity to reduce cooling times in the die and subsequent die spraying times and therefore, potentially higher productivity.

SSR, by introducing the extra step involving in situ slurry generation, will add operational cost and complexity to the die casting process; therefore, at the very least, the increased productivity value must outweigh this corresponding increase in cost. Castings that would be improved through a combination of the above mentioned factors are typically pressure-tight sensitive castings. SSR has been evaluated in several locations and is apparently in production of at least one.

However, since the slurry is made at the machine, each shot seems to be a unique event and process control presents a challenge. It is also unclear how the issue of oxides and contaminants may be handled on a production basis, and these issues may continue to restrict slurry based systems to pressure tight applications, of which, of course, there are many.

10.7 Other Slurry Based systems

10.7.1 "Semisolid" Slugs from Slurry

Some proposed alternative approaches to production of slurry on-site include generation of "semisolid" slugs. The basic technology is similar in many respects to the rapid cooling technology mentioned in the section on billet production. This approach builds upon the tendency of a highly grain refined alloy to form very fine equiaxed dendritic microstructures, which then coarsen rapidly to form spheroids while held between the solidus and liquidus temperatures. In one approach, a carousel is used to sequentially collect, cool and coarsen the alloy prior to a final induction stage of "equilibration" just before transferring the "slug" to the casting machine. In another, the actual shot sleeve of a vertical casting machine is used to achieve the same objective. It is believed that at least one such machine is producing parts commercially.

Finally, recent work has been carried out to marry the carousel vertical induction slug heating approach to an electromagnetic stirrer, such that a producer may select whichever process is best suited for a particular application. If slug-based technologies were appropriate, the system would be used conventionally. If raw material cost is an issue, the induction heating coils can be replaced by an electromagnetic stirrer, which would stir melted alloy delivered from a holding furnace prior to a coarsening stage at subsequent stations in which a virtual slug would be formed. After coarsening, the virtual slug is transferred to the casting cell.

This last approach has already been demonstrated on a pilot scale for both aluminium and copper based alloys with good success.

10.7.2 Direct Slurry Forming

Another process for producing semisolid feed material is the so-called direct slurry forming (DSF) process, which involves the bulk production and holding of semisolid slurry in a charging vessel [6].

Gibbs Die Casting Corp., a major automotive aluminium die caster, have developed a process capable of at least a 550 kg h^{-1} production rate, which also accommodate the vacuum ladling process Gibbs used for its die casting operations.

The Gibbs process can be separated into three steps: formation of semisolid metal, maintenance of the slurry suspension and transport of the semisolid mixture to the die casting machine.

Die casting operations typically have furnaces for melting ingot, and the molten alloy is transported to holding furnaces at the casting press. This metal can be as hot as 650°C at the holding furnaces. Producing 550 kg h^{-1} of partially solidified slurry is a matter of controlled heat removal; incoming molten metal must be cooled and stirred sufficiently fast to remove enough heat to lower the temperature to the desired slurry temperature – typically about 600°C corresponding to a fraction solid between 20 and 30% maximum to maintain fluidity for vacuum transport.

In this arrangement, where heat is removed through the walls of the furnace, dendrites tend to form at the outer walls of the furnace and grow inward as solidification progresses. Mechanically scraping the walls of the furnace at regular intervals during solidification breaks growing dendrites from the wall and moves them into the bulk of the bath. With sufficient shearing action, these dendritic structures are moved through the melt and coarsened into rounded structures by the flow of liquid around them. The inset in Fig. 10.12 illustrates the viscous consistency of the resulting slurry.

To overcome vertical segregation, an impellor with inclined rotating blades forces the flow of material upward in the melt. The final mixing system in the furnace consisted of an anchor-shaped rotor with vertical shearing rods located close to the furnace walls for the scraping of solidifying material, and an auger located within the rotor shape to promote vertical homogeneity. Figure 10.12 depicts the furnace and mixing configuration.

Figure 10.12 also shows the transfer tube, which is connected to a cold chamber shot sleeve. It was also found that temperature control of the ladle tube is especially

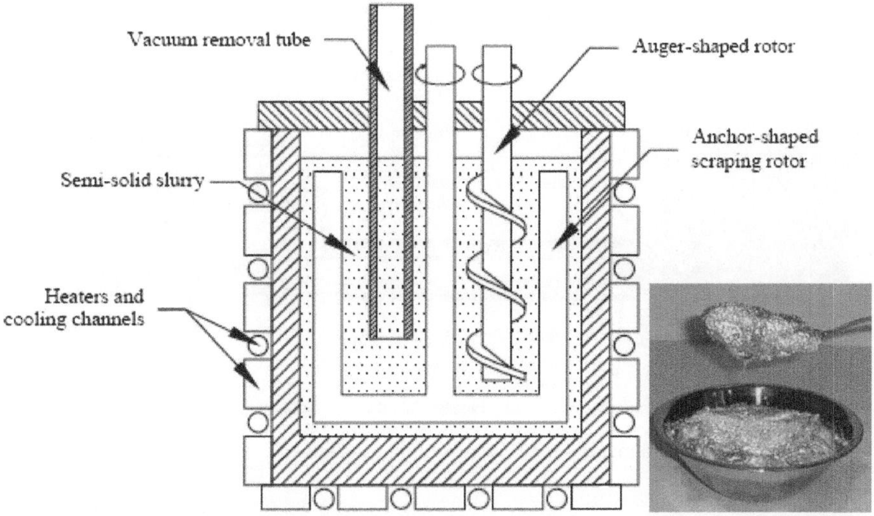

Fig. 10.12 Depicts the furnace and mixing configuration of DSF process with an anchor-shaped rotor with vertical shearing rods located close to the furnace walls for the scraping of solidifying material and vacuum drawing liquid aluminium through a tube into the shot sleeve of a cold chamber die casting press and photo insert the consistency of the slurry Reprinted with permission of ASM International(r). All rights reserved. www.asminternational.org

important in the case of semisolid material to prevent freezing of the slurry. The tube must be kept at or above the liquidus temperature of the alloy.

Each of these slurry-based approaches offer the attractive advantage of eliminating the premium billet cost associated with Billet, or Thixo-based systems. However, while each claims to reduce capital equipment cost, it is not clear whether this has, in fact, been achieved. Nor is it clear or demonstrated commercially that the added processing costs associated with tending these slurry makers do not offset the cost advantage over billet. For example, the costs associated with melting, treating and holding liquid metal are an added cost relative to a thixo-based operation although many of these slurry-based attempts are in fact grafted onto existing foundry operations where the necessary infrastructure is already in place.

10.7.3 Thixomoulding

Thixomoulding [7] is the injection moulding of thixotropic metal alloys (magnesium to date) in a semisolid or plastic-like state. The process, developed originally by Dow Chemical, one of the first investors in semisolid technology and an original licensee of the MIT patents, takes place in a specially designed machine resembling a screw driven, plastic injection-moulding machine.

Magnesium casting has always been a difficult exercise because of the natural flammability of the alloys in contact with air. The potential to process magnesium

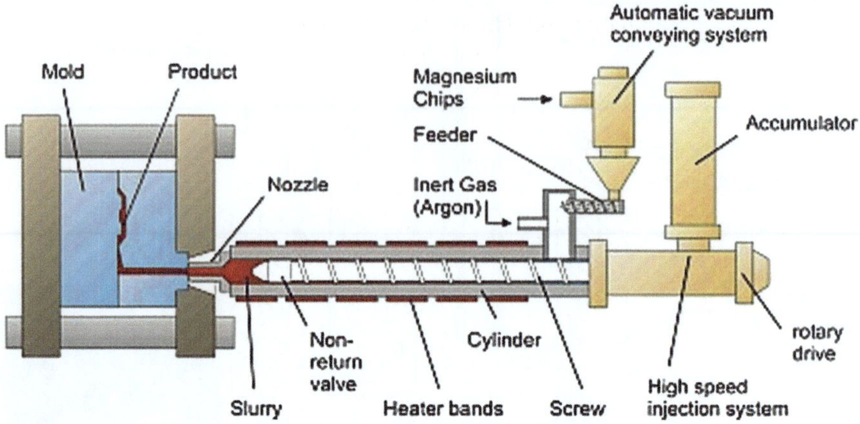

Fig. 10.13 Schematic of a magnesium thixomoulding machine [7]

in a machine akin to plastic injection moulding offers an opportunity to eliminate liquid magnesium from the factory floor and therefore, greatly mitigate these risks. Furthermore, magnesium offers an additional advantage, not available to aluminium, in that it has low reactivity in relation to ferrous alloys. It can, therefore, be processed into the semisolid state contained within steel devices (Fig. 10.13).

In the Thixomoulding process, chipped alloy magnesium is loaded at room temperature into a hopper on top of the machine. The granules are volumetrically loaded into a smaller hopper that mounts directly to the barrel. A rotating screw meters the material along the length of the barrel. The screw rotation produces a shearing action which together with heat applied externally forms a thixotropic semisolid slurry. The material, in the form of a low solid content slurry ($F_s \sim 0.05$–0.3 maximum), is then forced past the shut-off valve and injected into a heated mould. Once the part solidifies, it is removed and trimmed.

Because the technology requires very tight control of temperature, it is claimed that improved dimensional tolerance can be obtained and also reduced shrinkage porosity can result from injecting semisolid material. Unlike in the aluminium case, however, the fractions solid typical of Thixomoulded parts are quite low as identified above and therefore, this attendant advantages of reduced shrinkage and improved viscous flow would also be reduced.

First proven as a commercially viable technology in the electronics and telecommunications sector, Thixomoulding is used today for cell phone and laptop computer covers, photo and video cameras and many other applications.

Thixomoulding worldwide currently produces several internal and external parts for mainframe computers, servers and other electronics and telecommunications applications as well hand-held power tool components

Thixomoulding offers an alternative to traditional magnesium die casting and provides an attractive auxiliary process to companies active in injection-moulded plastics. For a plastic injection moulder, competition against magnesium die castings has been an ongoing battle and Thixomoulding presents a process closely similar to a plastic injection moulders core technology. Thixomoulding offers indeed an opportunity too for a plastic injection moulder to beat the magnesium die casters at their own game as it were.

However, despite wide adoption in Asia with more than 200 installations claimed, it has not caught on in either Europe or North America and, where it has, has been mostly at established plastic injection moulding houses. To date, no magnesium die caster has been identified as switching to Thixomoulding technology, and this is perhaps also because a special raw material is required, this time in the form of magnesium pellets.

10.7.4 Rheo-Die casting Process

A close variant of thixomoulding is the rheo-die casting process, which is an innovative one-step SSM processing technique to manufacture near-net shape components of high integrity directly from liquid Mg- and Al-alloys. The process innovatively adapts the well-established high shear dispersive mixing action of the twin-screw extruder (originally developed for polymer processing) to the task of in situ creation of SSM slurry with fine and spherical solid particles followed by direct shaping of the SSM slurry into a near-net shape component using the existing cold chamber die casting process.

Figure 10.14 schematically illustrates the rheo-die casting equipment for Al-alloys. It consists of two basic functional units, a twin-screw slurry maker (the key technology) and a standard cold chamber HPDC machine. The rheo-die casting process starts from feeding predetermined dose of liquid metal from the melting furnace into the slurry maker where it is rapidly cooled to the SSM processing temperature

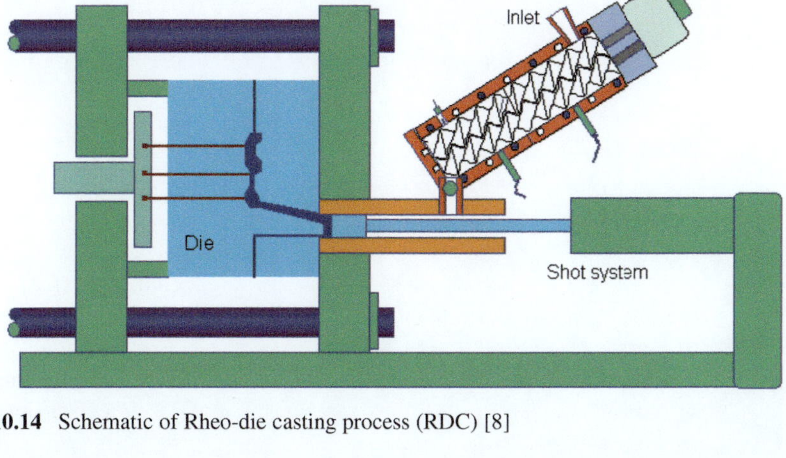

Fig. 10.14 Schematic of Rheo-die casting process (RDC) [8]

while being mechanically sheared by a pair of closely intermeshing screws converting the liquid into a semisolid slurry with a pre-determined volume fraction of the solid phase dictated by the barrel temperature. The semisolid slurry is then transferred to the shot chamber of the HPDC machine for component shaping. In order to prevent Al-alloy from oxidation, nitrogen gas is used as the protective environment during the slurry-making process. The fluid flow inside the twin-screw slurry maker is characterized by high shear rate, high intensity of turbulence and cyclic variation of shear rate. Although no information is publicly available on the material used for the screw mechanism, concerns are expressed as to the possible problem of high wear rates due to the high reactivity of liquid aluminium.

Chapter 11
Process Control in Die Filling and Die Design

Until recently, much of the SSM technology was highly proprietary and unknown to the general casting world. In the last several years, however, sufficient information has become available to make it possible to identify a number of alternative technical solutions that have been implemented around the world for the commercial production of high quality parts.

Of particular note, it is now clear that real-time controlled machines are not an absolute pre-requisite for the production of safety critical or other high quality parts. Many parts are being produced in high volume and with excellent results on standard three-phase injection machines ((a) pre-set injection velocity, (b) pre-set ramp and (c) pre-set consolidation time under pressure before final ejection), as indeed were the very first semisolid die castings produced in the MIT laboratories. Rather, the key ingredients of a successful SSM system approach are:

- A reliable source of consistent quality raw material
- An appropriate delivery system for semisolid material to the casting machine
- A diaphragm or alternative approach to strip oxides from the slug's surface or delivery system and eliminate or minimise the entrapment of oxides within the formed parts (see Figure 11.1)
- A powerful, repeatable, injection system capable of generating sufficient static pressure to feed solidification shrinkage throughout the freezing cycle
- Appropriate processes and controls to optimise the heat-treatability of the formed parts

Coupled with intelligent die design to facilitate both turbulent-free filling and adequate feeding of shrinkage, these features have a proven record of producing high quality parts over long time periods.

In general, die design must allow for relatively massive gate and runners systems, so that die filling can be accomplished quickly, without turbulence and the resulting gas entrapment, which is the death knell for full T6 heat-treatability. Entrapped gas, which is at high pressure after casting, expands upon heating and creates blisters as the metal softens during the high temperature portion of T6 heat treatment. Flow velocities can be higher than in squeeze-casting, since the semisolid alloy has a high apparent viscosity relative to fully liquid metal, reducing turbulence, and this allows SSM casting to produce wall thicknesses, which can be less than 2 mm.

David H. Kirkwood et al., *Semi-solid Processing of Alloys*,
Springer Series in Materials Science,
DOI 10.1007/978-3-642-00706-4_11,

Fig. 11.1 Representative diaphragm style to minimise slug surface oxide entering the die cavity [10]

However, typical plunger velocities around 25–50 mm s^{-1} and typical metal velocities of around 1–2 m s^{-1} remain at least one order of magnitude less than conventional die casting.

As a second criterion, the gate and runner must also be sufficiently thick to allow feeding throughout the part freezing cycle. In practice, this means that the gate and runner thickness must be typically >0.90 of the part thickness to be fed, although the metal flow through the gate tends to supply heat to the gate and runner regions during filling. This is an important criterion, since not all parts have massive sections and therefore, SSM parts do not always require thick gates.

11.1 Die Temperature Control Systems

As in any good casting process, good overall process control is an essential requirement for consistent quality. A key element in any die casting process is the die temperature, and modern die cast foundries invariably use dies pre-heated before casting and internally cooled during a run. The dies used during a recent Al SSM study [10] were heated either by re-circulating hot oil though internal passageways or using electrical fire-rods inserted into straight holes drilled through the cavity and/or die base. For hot oil heating, two 24 kW oil units were used. The target die temperature for all studies was 200°C, a common die temperature for aluminium SSM, although recent reports suggest that even higher temperatures can be beneficial for thin wall SSM parts.

An advantage of electrical fire-rods for heating is that the initial capital investment is quite low and there is less restriction on maximum operating temperature. However, care is necessary to avoid premature short-circuiting within the wiring and loss of heating caused by abuse or simply via careless mould spraying. There is also a safety hazard associated with high voltage connections in the vicinity of the die area, although these issues can usually be mitigated by careful placement and shielding of the leads and connectors.

Electrical die heating has been a standard in at least two high volume fuel rail production facilities for almost 10 years. However, electrical heating also raises the potential that local hotspots can occur within the die close to the fire-rods.

Even when the heating is well controlled by thermocouples within the die, without limits on power, the fire-rods can reach 1,000°C during heating cycles. This local overheating can burn off ejector pin lubricant and cause seizing and catastrophic failures. Electric heating means in addition, that die cooling must be effected externally. Re-circulating hot oil avoids these issues and with current systems the oil can be heated to almost 300°C. In this study, the oil passageways were designed so that they could accommodate both oil and/or electric heating. That is, they were terminated with NPT threads and reamed to precise diameter to accept the fire-rods. Reamed-to-size holes are critical in the case of fire-rods, since oversize holes can cause overheating of the fire-rod due to lack of good contact with the die, which can lead to premature burn-out. Oil systems, however, require only rough sized holes. Oil systems also have the added advantage that they can also function as internal heat-sinks once operations have begun.

Internal water-to-oil cooling circuits within the oil system can rapidly decrease the oil temperature that then functions as a heat sink flowing through the die.

In this study, both electric fire-rod heating and oil heating were used to maintain die temperature. Notwithstanding the above considerations, both systems provided adequate die pre-heating, and casting results were similar in both cases.

11.2 Parts Selection for Formability Evaluation

Figure 11.2 shows a range of parts weighing from 240 g to around 1 kg, which have been used for one particular formability evaluation. Both parts commercially formed from alloy 357 and welded and heat treated by the customer were shown in Fig. 11.2a, b. The part shown in Fig. 11.2c was formed from alloy 319 and T-6 heat-treated. The part shown in Fig. 11.2a used a one-cavity die and the parts in Fig. 11.2b, c were two-cavity dies. Figure 11.2a shows a motorcycle swing arm mount, which also doubled as an ATV frame part was also selected for cosmetic validation, because it had a large, relatively flat surface that lent itself to polishing. While this actual part is not itself a decorative finish part, its large flat areas were chosen to be useful in revealing polishing and plating defects.

11.3 Injection System Evaluation

Parts were produced on machines that could be operated both in real-time controlled machine mode and with standard three-phase injection profiles. The 800 ton machine used in these trials also allowed a variation of three-phase control comprising five separate velocities to be implemented in a step-wise mode during injection.

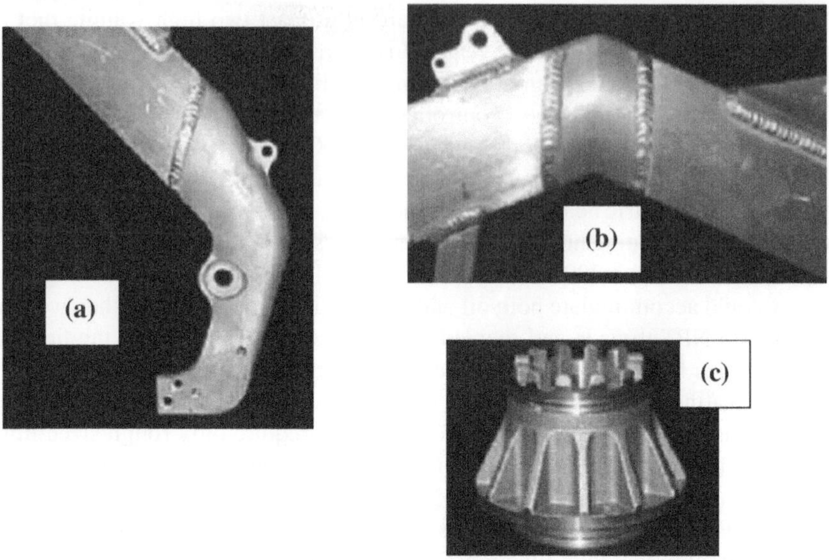

Fig. 11.2 (**a**) Motorcycle swing arm mount, alloy 357, welded to 6,061 extrusions and heat treated by the customer, one-cavity die, (**b**) extrusion connector ATV frame, alloy 357, welded to 6,061 extrusions and heat treated by the customer and (**c**) flange for demolition hammer, alloy 319-T6 heat treated, two-cavity dies

In all events, whenever engaged, real-time control mode only operated in the velocity control phase. Intensification during all injection cycles was triggered using standard techniques either based upon plunger position or on hydraulic pressure.

Pressure could not be modulated during solidification. Real-time pressure control has limited applicability at best, and was not considered desirable for production of any of the selected parts.

11.4 Fill Test Observations

Interrupted shots (fill-test) of the filling process for a motorcycle swing-arm are shown in Fig. 11.3. This die requires a 75 mm diameter slug, which when heated to the SSM condition, needs an 80 mm diameter shot sleeve. Plunger velocity during filling was established from production experience at approximately 250 mm s^{-1}; below this, the slurry freezes prematurely leading to incomplete filling and above it, leads to breaks in the flow front and associated turbulence. It should be noted that this velocity is roughly two orders of magnitude less than typical for die casting. For the die and shot sleeve configuration used in this study, this plunger velocity generates a metal velocity at the gate of approximately 1.6 m s^{-1}.

The fill-test programme allows the control system to stop the actual injection stroke at specified positions. In this way, especially at SSM velocities, the resulting

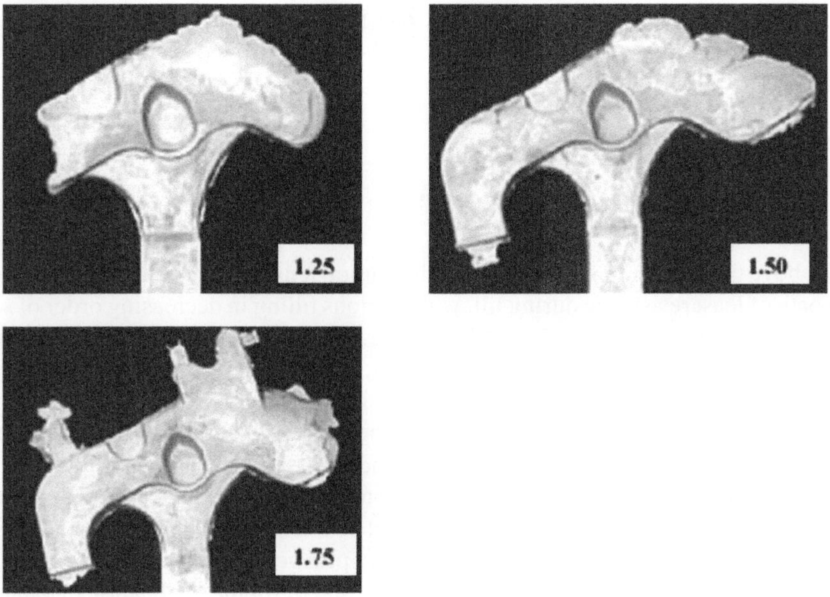

Fig. 11.3 Fill test of the motorcycle swing arm mount showing 1.25%, 1.50% and 1.75% filling of the die cavity

partially-filled cavity provides a snapshot of the metal disposition during fill. In conventional die casting, at much higher velocities and with more fluid material, more care is required in interpretation of this exercise, since inertial effects can still cause significant metal motion even after the plunger stops travelling.

Figure 11.3 confirms that the SSM material fills the cavity in essentially plane front mode flowing evenly from the gate toward the cavity extremities. This is a unique feature of SSM and allows venting, and particularly vacuum systems, to be correctly positioned for maximum utility. Of particular note for this study [10], this same filling pattern was obtained when the machines operated in either real-time velocity control or three-phase control. While the three-phase, three modes of operation resulted in an earlier deceleration of the plunger as the metal resistance built up, both real-time and three-phase operation gave identical results. This confirms that for most parts, the impact of real-time control is essentially insignificant. Indeed, for machines that are totally real-time controlled (velocity and pressure), the switch over to pressure control at the end of the stroke can lead to undesirable velocity spurts just as the filling process is completing.

It appears that the planar filling behaviour is much more a characteristic of the higher viscosity material than any special machine control feature. It should perhaps be remembered that the early pioneering work on the SSM process establishing the key advantages was also performed on a standard three-phase machine; so this result should not be too surprising. Similar fill patterns were obtained for the other parts selected in this study [10]. The smooth filling pattern is confirmed for each of these parts in production by a very low reject rate resulting from blisters after T-6 heat treatment, regardless of which machine, real-time controlled or three-phase, they were produced on.

11.5 Venting and Vacuum Assist

The above fill-test studies also illustrate the natural tendency for SSM flow to follow the path of least resistance during fill, with sections filling in decreasing order of section thickness and the thinnest sections filling last. This can be influenced, to some extent, by die temperature and/or die spraying, in that hotter die surfaces do encourage earlier filling. However, in general, considerations of solidification processes restrict the degree to which temperature can be used in this regard. The regular and predictable mould filling is both a blessing and a curse. It is usually quite simple to predict the last places to fill, and thereby locate vents. However, if thin sections are also associated with blind holes, which cannot be adequately vented and since SSM invariably implies plane front filling, increasing plunger velocity as in die casting, is normally not an option. Such situations are well suited for the application of vacuum to evacuate cavity gases ahead of the filling front.

The part shown in Fig. 11.5b represents a typical case story for vacuum assist. Each of the weld fitment areas contains thin, blind pockets and is troublesome to fill. The same is also true for the end of the swing arm mount shown in Fig. 11.2a. A 5.6 kW vacuum pump, in conjunction with a 113.55L vacuum receiver tank, was connected to the die. The system is shown in Fig. 11.4a, while Fig. 11.4b illustrates the vent path in the die from the part to the vacuum runner to the vacuum valve. Although SSM flow is so slow relative to conventional die casting, the same concerns exist about metal reaching the valves before closing, so a special mounting

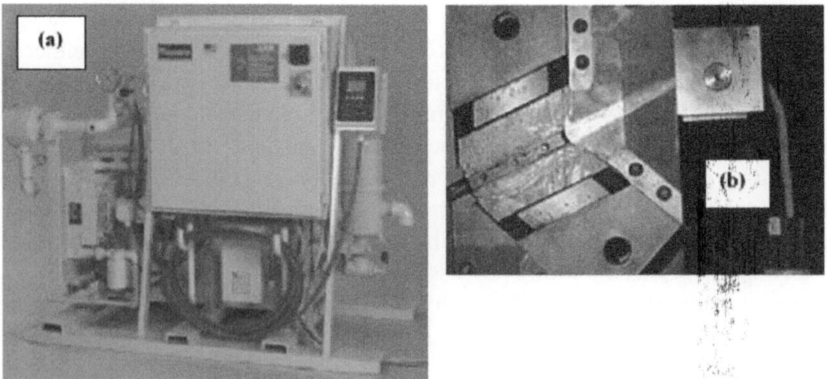

Fig. 11.4 (**a**) Die vacuum system with 7.5 HP pump and113.55L tank and (**b**) Vacuum runner for the part shown in Figure 11.5

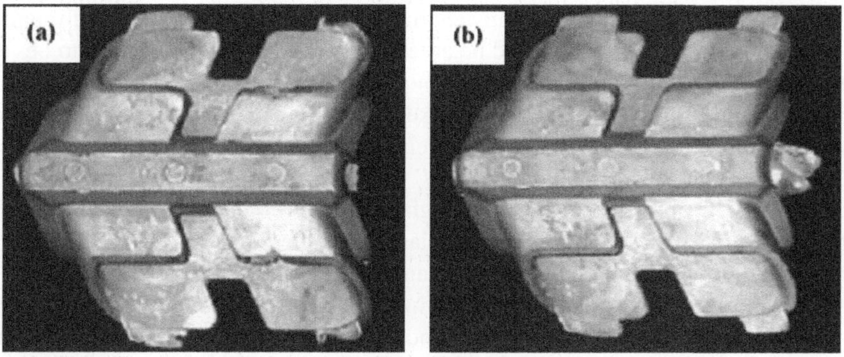

Fig. 11.5 Extrusion connectors produced without vacuum assist. Note *non-fill* areas (**a**) and Extrusion connectors produced with vacuum assist. Note *all areas* are now well filled

system was chosen to expedite the testing. It is possible, however, to leave the vacuum valve open and drawing vacuum until the cavity is nearly filled, because metal does not spray ahead of the flow front. In normal die casting, the valve would typically close as metal begins flowing through the gate.

Figure 11.5a shows a part made without vacuum and Fig. 11.5b shows a part made with vacuum. It can be seen that without vacuum the metal does not consistently fill the blind pockets. However, with the vacuum assist operating, the parts completely filled in the deep pockets, and the rejection from blisters due to trapped gas during heat treatment was also reduced.

11.6 Runner, Gate Design and Injection Forces

As already discussed earlier, the filling behaviour of SSM parts is very different from normal die casting. The material fills the cavity in essentially plane front, without turbulence and generally takes the path of least resistance filling thicker and/or hotter cross-sections before thinner and/or colder cross-sections. That is, to say that die temperature can be used to influence the mode of filling to some extent; although typically, it is more important to have a uniform die temperature than to adjust filling behaviour.

Blind holes, as already mentioned, can be troublesome. It is important to ensure that blind holes can be filled by venting cavity gases ahead of the filling front. Here, vacuum can play a critical role as already illustrated. In addition, the typical objective of SSM casting is to produce fully dense, heat treatable parts. Therefore, it is normally the case that the runner and gating design should allow for controlled freezing back to the gate. This means that not only should the gate cross-section be adequate to fill the part with a plane front, it should also be thick enough to freeze only after the part is frozen. This is quite different than for normal die casting and typically means that gates are so thick that they cannot be trimmed, but rather must be saw-cut from the part as in permanent mould casting. A typical "rule of thumb" is that the gate cross-section should be at least 90% of the casting section to be fed with liquid slurry during freezing. This is not necessarily the thickest section, which can be isolated from the gate, but the thickest well-connected section of the casting where "feeding" might reasonably be anticipated.

It is the requirement for this "feeding" that cavity pressures in SSM are often around 100 MPa or even higher. These higher cavity pressures as well as the inherently higher resistance to fluid flow associated with the SSM material are the basis for the higher power shot end typically found in SSM cells.

For example, the 400 ton system used in these trials had a modified shot end to provide up to 65 tons of injection force, and the 800 ton system with approximately 135 tons of injection force. These forces are far higher (almost two times) than the typical injection force, which might generally be 10% of clamping force on many die casting machines. This factor also reduces the largest part that may be made on a particular machine, since the clamping force applied to the die halves must be sufficient to withstand the opening force created by the instantaneous metallostatic pressure of injection multiplied by the projected area of the formed part. With higher injection forces, the metallostatic pressure is comparably higher and therefore, the largest part size reduced. Fortunately, most SSM parts are thick-walled structural or pressure type applications rather than typical thin-walled die castings.

11.7 Surface Quality

The swing-arm mount was also selected for this study, since it had a large, essentially flat surface appropriate for polishing. High lustre polish and/or chrome plating is an additional advantage of SSM processing, since SSM parts do not exhibit subsurface porosity, which can cause polishing and/or plating defects. Figure 11.6a shows a polished swing-arm mount and Fig. 11.6b shows a chrome-plated motorcycle part.

The surface quality is particularly good and reject rates are low when SSM parts are treated using these techniques. This advantage also applies to internal parts that may be plated for corrosion protection. For example, many electronic housings are plated for protection, and porosity, which leads to bleed-out of plating fluid after finishing, is particularly troublesome. As SSM casting becomes economical for copper

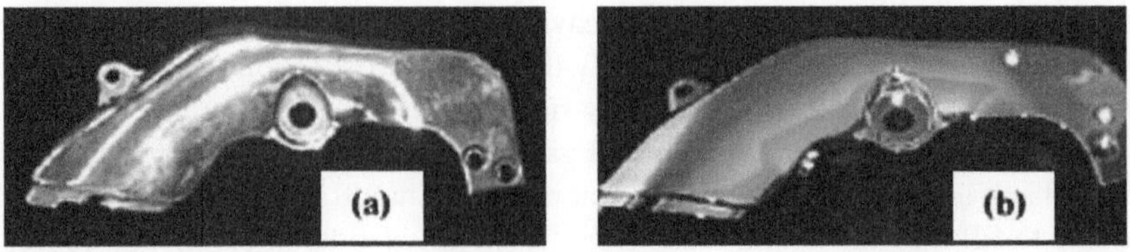

Fig. 11.6 (a) Polished swing arm mount from Figure 4.21 (a), and (b) Chromium plated SSM cast part.

alloy casting, this advantage will allow die casting to capture even more market share of plumbing, decorative hardware and lock components.

11.8 Summary of General Process Issues.

11.8.1 Raw Materials

Both electromagnetically stirred and rapidly chilled materials (no stirring) appear to deliver acceptable material using both horizontal and vertical casting systems. Recently, small-scale horizontal continuous casting systems have become available, which allow producers the opportunity both to recycle scrap and/or produce virgin SSM billet on-site, prepared from melted alloy right at the machine and just prior to casting (see Chap. 9), in addition to the so-called slurry on demand systems discussed in Sect. 9.6. Both these approaches address directly the high cost of raw material imposed by the primary suppliers and should, over time, force a downward trend in SSM raw material pricing.

11.8.2 Alloy Selection

The primary alloy used in today's SSM processing is aluminium alloy 357, which has been electro-magnetically stirred using radial stirrers while being continuously cast into round cross-section bars. This is the most economical method of producing electro-magnetically stirred bars and provides a consistent material cast over long periods of time from a small group of moulds. However, as mentioned earlier, more and more interest is being generated for slurry-based processes, which obviate the need for a continuously cast feedstock.

Chapter 12
Component Design Rules

12.1 Tolerances

Semisolid Metal Processes generate near net-shape end products with a high degree of precision because they use hard steel tooling and high injection pressures. This combination results in a high level of replication of the die contours and generally the NADCA standards for "precision tolerances" can be achieved on a regular basis.

12.2 Part Design Rules

It is important whenever possible to eliminate sharp corners and provide proper draft and radii that assist the smooth and complete filling of the die cavity under speeds associated with high production cycles. Such provisions will tend to extend die life as well as allow the manufacturers to make use of the natural advantages of the SSM processes.

Wall thicknesses typically range down as low as 1 mm, depending on the alloy system used, part size and shape and of course the intended application.

Apart from the linear dimensional tolerances, other tolerances will be additive, such as "Parting line, Moving die components, Angularity, Concentricity, Parting line shift, Draft requirements, Flat requirements, Cored holes for cut threads, Cored holes for formed threads, Cored holes for pipe threads, and Machining stock allowance" tolerances. The NADCA publication no: 403 on "Product specification standards for die castings produced by the semisolid and squeeze casting processes" provides a full list of tolerances for a variety of alloys systems [NADCA].

12.3 Dimensioning

The type of dimensioning chosen can have a significant impact on the ease or difficulty in establishing tolerances and specifications. There are two basic dimensioning methods: (1) the coordinate and (2) the geometric dimensioning and tolerancing

David H. Kirkwood et al., *Semi-solid Processing of Alloys*,
Springer Series in Materials Science,
DOI 10.1007/978-3-642-00706-4_12, © Springer-Verlag Berlin Heidelberg 2010

(GD&T). The former method defines parts by their location on a 3D grid is cumbersome and may not take into consideration the actual function of the part. In contrast, with the GD&T system, the entire part must be displayed, allowing the interrelationship between the orientation and location of features to be displayed. Therefore, the GD&T system specifies form and positioning tolerances based on the interaction of part features and therefore, is more flexible.

12.4 Cost Reductions in Product Design

The two basic geometric factors that play a major part of production costs are:

1. Design configuration
2. Specified tolerances

Geometric tolerancing defines a part in accordance with its function; functional dimensioning. Functional dimensioning can often result in larger manufacturing tolerances that meet the precise functional specifications of the part, thus resulting in cost reductions. A good design will aim to provide the largest possible production tolerances that are compatible with the specified functionality. In the harsh competitive environment of today's globalized manufacturing, "a larger tolerance is less expensive to produce than a tighter one" is not an understatement, but a universal truth.

Design costs are the major component in the making of any product. They are estimated to be as high as 80% of the total manufacturing costs. Decisions taken in the design stage are carried through the whole manufacturing cycle in the life of a product, i.e., tooling, production, gauging, inspection, etc. Designers are not only responsible for producing a design that caters for how the part is made, but also as to how the part functions and these must operate within the constraints of economically viable production. As in everything else in life, design requires choices and therefore compromises, what is called "optimization through balance." A good design will optimize the compromises between part function and part cost, thus achieving the necessary balance that will promote and ensure product success.

Chapter 13
Practical Applications in Use Today

Prediction is difficult, especially about the future, said J.D. Barrow; amusing and very true. However, there is no doubt that in today's manufacturing business world one cannot avoid the fact that the rate of change is accelerating. Apart from conventional incremental innovation, radical innovation has become strategically very important and used as a source of competitive advantage in the global market place.

The development of semisolid metal processing is an example of the full spectrum of experiences involved in turning a "research idea" to a successful "commercial product".

Following an unsuccessful attempt to exploit the technology directly by forming an entrepreneurial company, located on the famous Route128 ring road around Boston, the original MIT technology patents eventually were licensed to International telephone and telegraph corporation (ITT Corp.) in the late 1970s. ITT Corporation held the technology very closely, a practise which was also adopted by Alumax Inc., which acquired the technology from ITT Corporation around 1985. The philosophy of both corporations was to maintain tight secrecy around all developments of semisolid processing and to resist request for licensing or joint venture without stringent demands, which effectively eliminated all opportunities for the type of cross-fertilization, which in turn so greatly evolves as technical innovation.

However, by around 1985, ITT Teves, a subsidiary of ITT Corporation and a leading manufacturer of braking systems, had established a semisolid manufacturing facility in Northern Germany, which sparked interest in a number of European manufacturers. As a consequence, in the late 1980s, a number of European manufacturers, having been rebuffed by Alumax, began to develop their own independent approaches for both raw material production and parts forming.

Buhler, a leading European die casting press manufacturer, began to promote their real-time controlled machines as suitable for semisolid processing and their partner at the time Alusuisse developed an electromagnetic stirring system and horizontal continuous caster to produce raw material. Similarly, Pechiney, in France (now a division of Alcan Aluminium) began their work on a vertical continuous caster, for raw material using a combination radial and longitudinal electromagnetic stirrer. Also in Italy, Magnetti Marelli developed a convective slug heating system in combination with machines from IdraPresse for the production of automotive fuel rails.

David H. Kirkwood et al., *Semi-solid Processing of Alloys*,
Springer Series in Materials Science,
DOI 10.1007/978-3-642-00706-4_13,

Of these European developers, by far the most aggressive proponent of semisolid processing was Buhler, who used the opportunity to increase the sales of their horizontal die casting machines. In relatively quick succession, Buhler assisted first Stampal S.p.A in Turin and then Hot Metal Molding Inc. in Arkansas, USA, to enter production of automotive components using raw material from either Pechiney or Alusuisse or one of the several "copycat" producers, who quickly ramped up raw material production. By the mid 1990s, Buhler had more than 40 machines installed around the World for the production of SSM parts.

Stampal S.p.A. in Turin invested in a number of horizontal Buhler machines and Fiat became a keen supporter of the process subsidizing some exciting innovative parts for its Punto, Panda and Alfa Romeo models (see Fig. 13.1).

However, the automotive market is exacting on price and before long, the premium price of raw material produced by the various electromagnetic casting systems was too much for the market to bear and many of the so-called billet based producers could no longer compete. It could be argued that in some instances it was largely because of poor choices of target parts, but at any rate, the interest has turned to slurry-based systems, which promise to eliminate all premiums and facilitate on-site recycling. By the late 1990s, UBE, a Japanese die casting machine manufacturer entered the market with the new rheocasting process (see Sect. 9.6).

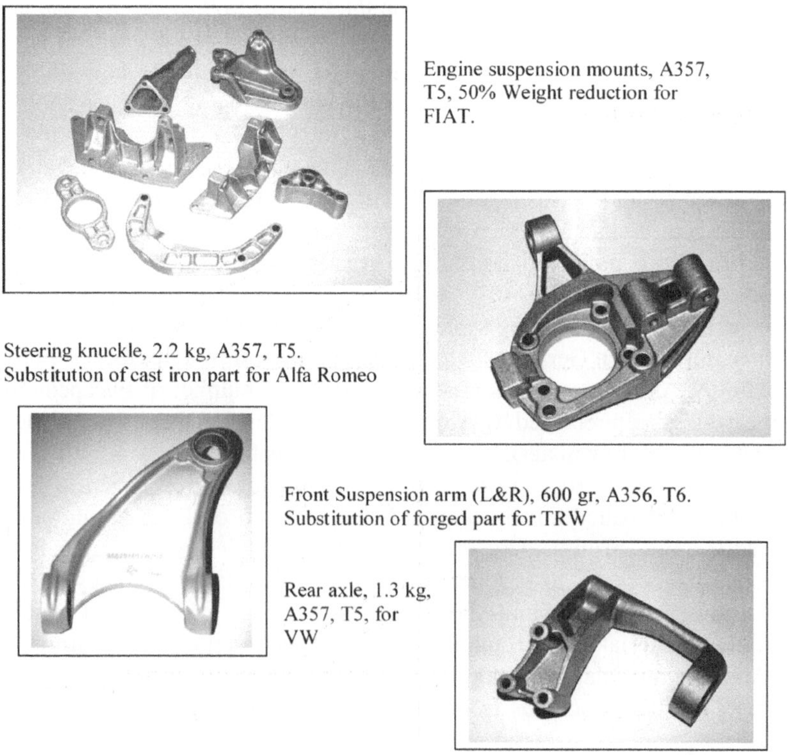

Fig. 13.1 Parts produced by Stampal S.p.A for various automotive manufacturers using the MHD feedstock route

Currently, more companies appear to develop parallel production routes that allow in-situ re-cycling, such as the SSR (Sect. 9.6.2) and other developments are on the way by the aluminium manufacturer Alcan [11] and the die casting machine producers Buhler. While UBE's new rheocasting technology grabbed a lot of attention in the opening years, commercial operations using this approach outside of Japan appear to be limited to Stampal S.p.A in Italy. Those installations originally imported in the USA are either mothballed or ceased entirely and semisolid processing is in a "wait-and-see" mind frame while several other slurry-based systems are evaluated. In the meantime, two significant producers continue to use billet-based technology in the US: Madison Kipp Corporation (Wisconsin) focusing mainly on automotive applications with four cells, Vforge (Colorado) focusing primarily on non-automotive applications, and SAG in Austria continuing to produce automotive products using its two SSM cells.

Semisolid metal Processing has become a widely accepted industrial manufacturing route for the production of near net-shape of aluminium and magnesium alloy components. Recently, there have been published quantitative data in support of the assertion of near net-shape. Being a relatively "new process", thixoforming had to establish itself by exploiting alloys that were already available. It would have been extremely difficult for thixoforming as a relative new process to establish itself in the market, if it required from industry to accept the use of new and unfamiliar alloys, as well as having to cope with the novelty of the manufacturing process itself. As a result, the widely tried and tested 356, 357 and A357 aluminium–silicon alloys were chosen as the materials, on which the process could demonstrate its potential. This phase is now over and millions of automotive parts are now in everyday use in the cars we drive [12].

It should be borne in mind that total production of aluminium castings is in the order of 1,500,000 m in North America and Europe and 1,000,000 m in Japan. However, thixoforming represents only around 1% of the total in this vast market where die casting and permanent mould casting are still overwhelmingly dominant.

In order to grow out of its current minor niche status, SSM processing has to overcome restrictions on three fronts:

- In dynamic applications for load bearing components (fatigue properties)
- In alloy development – expanding the existing portfolio of thixoformable alloys (better mechanical properties)
- In materials recycling

Although the position of the MHD route of feedstock production had been virtually unchallenged throughout the development of the thixoforming process, its inherent drawback of non-direct recyclability, involving added costs, has attracted the main challenge from NRC, and now from the other rheocasting variants discussed above.

In parallel with all these activities, Thixomoulding, the injection moulding of thixotropic metal alloys (magnesium to date) in a semisolid or plastic-like state, developed by Dow Chemical (see Sect. 10.6), is used today for cell phone and laptop computer covers, photo and video cameras and many other applications (Figs. 13.2–13.15).

Fig. 13.2 220t-Thixomoulding®-machine from Japan Steel Works at Neue Materialien Fürth GmbH (NMF)

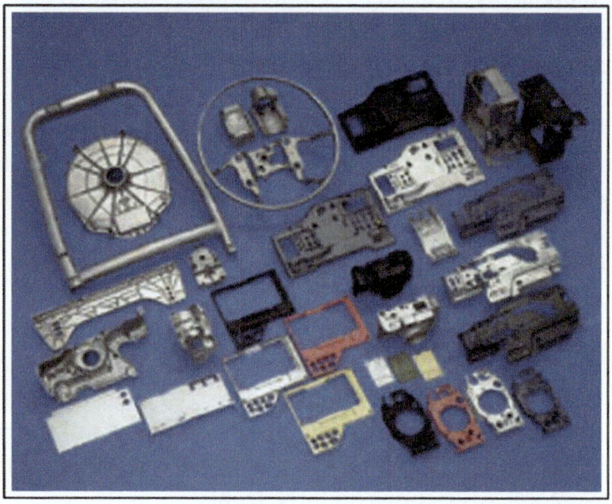

Fig. 13.3 Examples of magnesium thixomoulded parts from JSW Inc

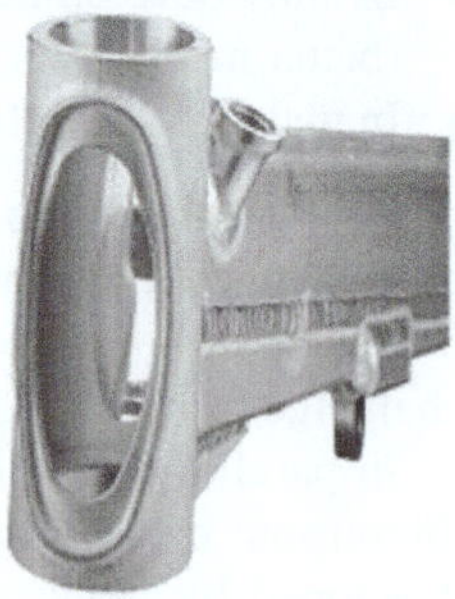

Fig. 13.4 Off-road motorcycle head-tube, alloy 357-T6 welded to a 6061 extrusion to make the complete frame (Vforge)

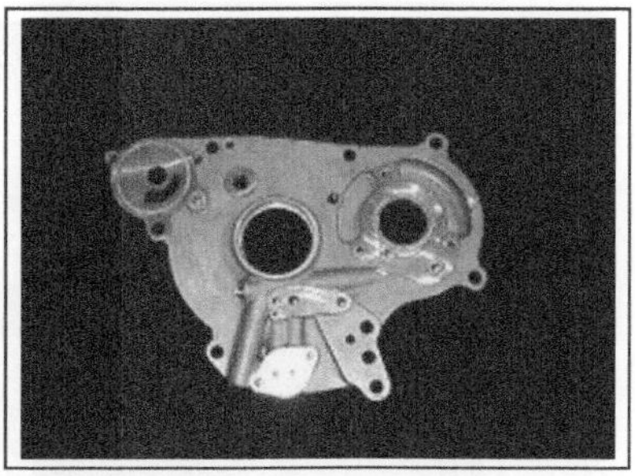

Fig. 13.5 Small gasoline engine cartridge plate formed from alloy 357-T5, which replaced a machined alloy 6061-T6 piece (Vforge)

Fig. 13.6 Small gasoline engine containing a number of SSM aluminium alloy parts including the entire crankcase, which is formed from alloy 357-T5 using a 2,500 ton machine and 5 in. diameter billet material. The cartridge plate in the previous figure and a companion SSM formed plate can be seen installed within the mechanism (Vforge)

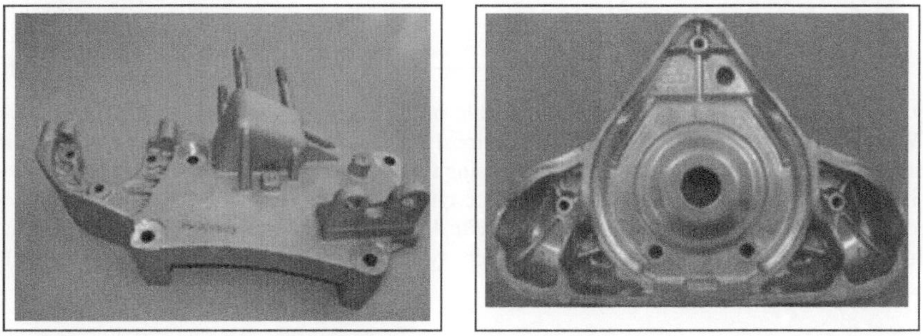

Fig. 13.7 Rheocast high strength safety parts: engine bracket (*left*) and Lagerbock (*right*) both using A 357 alloys (courtesy of Stampal S.p.A)

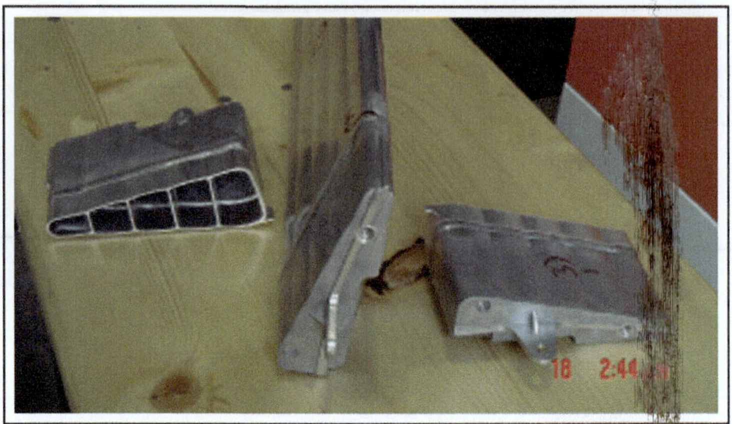

Fig. 13.8 Air tank side parts for extruded tank profile; operating pressure 16 bar, very thin walls, 240 g each part using the MHD route. Material: New alloy, high YS (290 MPa) and E 12 (%) no H.T, low porosity, weldable, Audi. (Courtesy of SAG GmbH)

Fig. 13.9 *Left*: Air manifold harness weldable on hydroformed part, thin wall, high ductility (VW and Audi) 130 g, replaces two parts previously produced by stamping. Front part of engine, vibration issues (fatigue); *Centre*: Antenna housing for tower mounted amplifier (TMA). Corrosion resistant, precision near net-shape, good surface quality, easy to coat (3–5 μm silver), with good electrical and inter-modulation properties; *Right*: Flange energy crash absorber; near net-shape component for AUDI A6 V8 energy management system for bumpers (Courtesy of SAG GmbH)

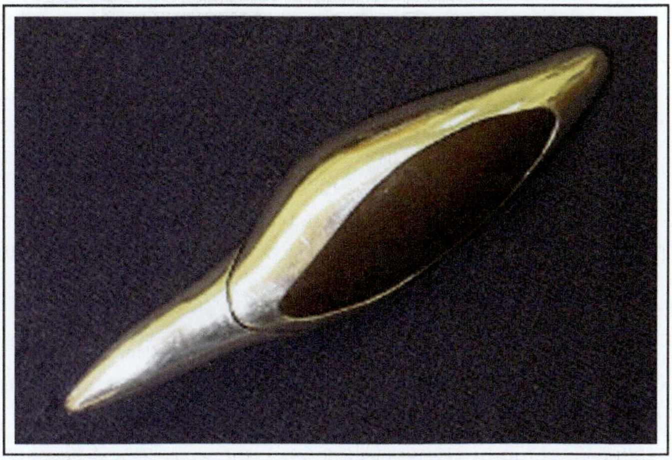

Fig. 13.10 Decorative pen and cover SSM formed from aluminium alloy 6061 in order to achieve a very high lustre polish. Aluminium alloy 357 polished to a low lustre finish because of the silicon content, and so this was one of the very first wrought aluminium alloy production applications (courtesy Vforge)

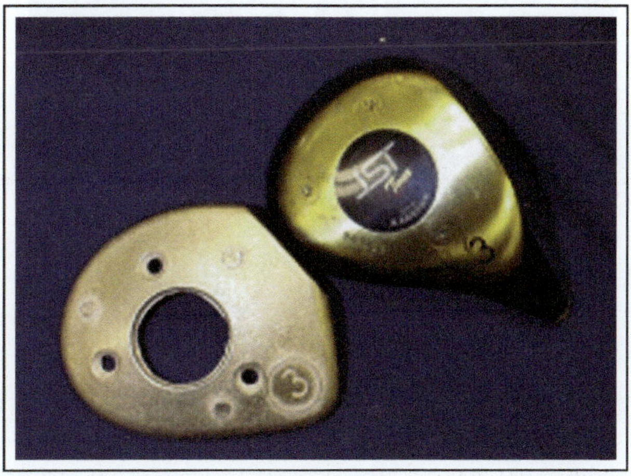

Fig. 13.11 Golf putter formed from alloy 357-T5 over a brass rod acting as a weight (Vforge)

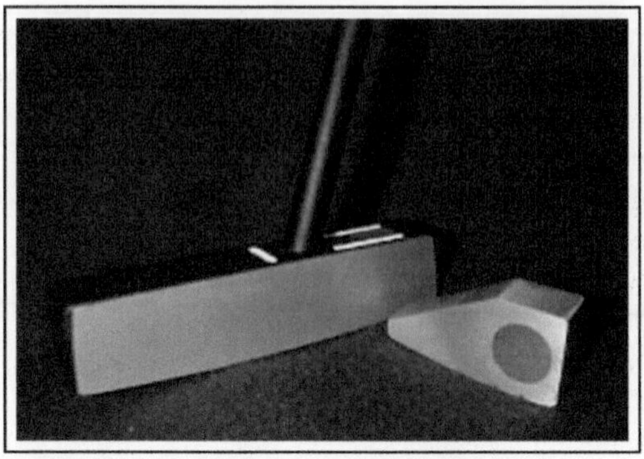

Fig. 13.12 One of the very first SSM copper alloy production parts. A golf driver sole plate formed from aluminium bronze alloy replacing a stamped and machined alloy 377 piece. This application offered improved wear resistance from the superior alloy as well as a near net-shape without machining requirements (Vforge)

Fig. 13.13 Section views of the oil pump filter housing produced by SSR (Courtesy Idra Prince)

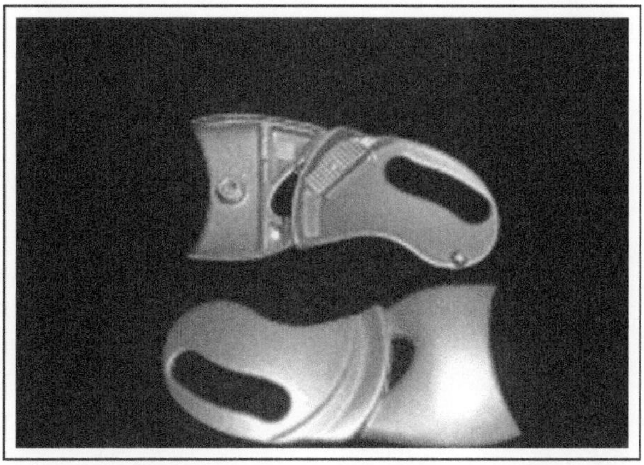

Fig. 13.14 Orthopedic knee joint piece replaced an investment cast steel piece saving 0.25 lb weight per knee! Alloy 357-T5 (Vforge)

Fig. 13.15 Some die cast dies can be converted to SSM forming provided the gating is acceptable for SSM die design as discussed above. This centre-gated die was converted relatively easily and runner volume saved by moving the shot position to dead centre (Vforge)

Chapter 14
The Future

As mentioned above, the premium cost of raw material has been an encumbrance to the widespread adoption of the technology. While niche producers, such as Vforge continue to grow their market share and specialised automotive products manufacturers, such as Madison Kipp Corp. continue to prosper, for semisolid metal processing to become once again a key player in the large-scale production of automotive components, a manageable solution to the cost of raw material must be demonstrated. Much is discussed currently within the field of various slurry-based approaches, but so far none are known to have made a serious inroad into automotive production where cost is as paramount as quality and performance.

In the meantime, slurry technology shows promise as a means to lower the cost of conventional die castings by providing a partially solid charge material for parts already designed as die castings and thereby accelerating production and improving quality, but once again the jury remains out as to the trade-off between cost of the technology (both capital cost and operating cost) versus the realised benefits.

In addition, although SSM of high melting point alloys offers exciting possibilities and tremendous potential, and has already been part of the original work of over 30 years ago [13], it is still at present in the research and development stage. More recent activities have produced complex shaped parts in steels as well as providing favourable direct economic comparisons with conventional forgings, as well as further development in ceramic or hybrid die materials [14]. These activities have now been consolidated under a European consortium with a brief to commercialise the thixoforming of high temperature alloys using a variety of steel grades as a starting point [15].

David H. Kirkwood et al., *Semi-solid Processing of Alloys*,
Springer Series in Materials Science,
DOI 10.1007/978-3-642-00706-4_14,

References for Part III Industrial Applications of Semisolid Processing

1. D.P. Spencer, R. Mehrabian, M.C. Flemings, Metall. Trans. **3**, 1925 (1972)
2. K.P. Young, C. Rice, in *Viscous, Semi-Solid Forging of Aluminum Reduces Costs and Boosts Alloy Flexibility, Aluminum USA*, Navy Pier, Chicago, USA, 2003
3. H. Kaufmann, H. Wabusseg, P.J. Uggowitzer, in *Metallurgical and Processing Aspects of the NRC Semi-Solid Casting Technology,* Aluminum, 76. Jg., 2000
4. J.A. Yurko, R.A. Martinez, M.C. Flemings, in *SSR: the Spheroidal Growth Route to Semi-Solid Forming,* Proceedings of the 8th International Conference on Semi-Solid Processing of Alloys and Composites, Limassol, Cyprus, 2004
5. M.C. Flemings, W.L. Johnson, *High Viscosity Liquid and Semi-Solid Metal Casting: Processes and Products*, Plenary Lecture, World Foundry Conference, Kyong Ju, Korea, 2002
6. C.S. Rice, P.F. Mendez, Adv. Mater. Process. **159**, 49 (2001)
7. *Thixomolding* – http://www.thixotech.com/thixomld.htm
8. X. Fang, J. Patel, Z. Fan, in *Microstructure and mechanical properties of rheodiecast (RDC) aluminum alloys, 67th World Foundry Conference*, Harrogate, 2006
9. A. Bendada1, C.Q. Zheng, N. Nardini, Phys. D Appl. Phys. **37**, 1137 (2004)
10. K. P. Young, C. Rice, B. Homiston, C. Nelson, in *Observations of fill behavior, surface quality, weldability and performance of SSM aluminum structural parts produced on real-time and three-phase machines*, 2001 Die Casting in the 21st Century, Ohio, Cincinnati, USA, 2001
11. D. Doutre, J. Langlais, S. Roy *The SEED Process for Semi-Solid Forming, Proceedings of the 8th International Conference on Semi-Solid Processing of Alloys and Composites*, Limassol, Cyprus, (2004)
12. P. Kapranos, Fonderia Pressofusione 28, (2007)
13. M.C. Flemings, R.G. Riek, K.P. Young, Mater. Sci. Eng. **25**, 103, (1976)
14. P. Kapranos, Chuzo kogaku (Chuzo kogaku), Source (Semi-solid processing), **77**(8), 518 (2005)
15. *Semi-Solid Processing of Steels –Thixosteel* http://www.cost.esf.org/index.php?id=943

Acknowledgements

The authors would like to express their thanks to the following companies for allowing us to use photographs of their SSM products: Idra Prince, JSW Inc, SAG GmbH, Stampal SpA and Vforge Inc.

Appendix to Industrial Applications of Semisolid Processing

Mechanical Properties of Some Thixoformed Alloys

D.H. Kirkwood and P. Kapranos

Commentary on the Mechanical Properties of Thixoformed Aluminium Alloys

The properties of the thixoformed alloys will depend on a large number of factors: on the exact chemical composition, including impurities such as oxygen pickup, on the method of manufacturing the raw billets (mechanical stirring, MHD, grain refinement, re-crystallisation, low superheat, etc.), the processing parameters involved in shaping the component (ram speed, die temperature, component dimensions) and the shaping process itself (rheocasting, thixocasting or thixoforging), all of which may affect the grain size and morphology; and finally, any subsequent heat treatment given to the product. Variations in the above factors will affect the measured property values, quite apart from measurement errors themselves.

The properties gathered from the literature reveal some variation as a consequence and in order to assess the important factors that need to be controlled, greater emphasis has been put on the variations in property values originating from the same source, where some factors (e.g. manufacture, composition, heat treatment) can be assumed to be constant, and from those sources with a long experience of producing the raw material under carefully controlled conditions, which generally means the industrial producers. The typical values of properties given are not averages, since the outlying values have not been included, but the modal values about which the most reliable values cluster.

Casting Alloys

Al–Si–Mg Alloys

We note firstly in Table 1 that in A356 (0.3 Mg%), the YS and UTS values are very consistent in both the T5 and T6 conditions. The exceptional low values probably arise from porosity [11], introduced from mechanical stirring and inadequate injection pressures. The exceptionally high values [12], however, are apparently obtained from specimens machined out of the as-received billets but not actually thixoformed, and therefore are essentially pore free. The noticeable variation in ductility in the T5 condition observed in the table maybe due to the variation in eutectic silicon morphology, see Uggowitzer [7], which may be spheroidised by a short high temperature treatment.

The strength of thixoformed A357 (0.5% Mg) is clearly increased by the addition of more magnesium, allowing a greater precipitation of Mg_2Si-the strengthening phase. However, it is apparent that the ductility falls, and increasing the magnesium to 0.8% in an experimental alloy [7] further increases the T6 strength with a fall in ductility. This has been shown to result from the formation of coarse clusters of silicon during long times at the high solution treatment temperature. A short high temperature treatment of 3 min, however, spheroidises the silicon without significant coarsening, rendering this alloy highly ductile in the T6 condition, without too much loss in strength. This treatment could be very effective in all high Si containing alloys.

Al–Si–Cu–Mg Alloys

C555 (Al6Si1CuMg) is designed to provide greater strength in the T5 condition than 357, with sufficient ductility [3].

319 (Al6Si3Cu 0.3Mg) provides the greatest strength of all the conventional casting alloys, achieving a UTS of 400 MPa, which in a modified condition, may develop a T6 yield strength of 350 MPa [16]; this alloy also shows impressive fatigue strength in this state. The tensile strength of this alloy is only exceeded by that of the exotic Al–Cu–Mg 201 alloy [18], containing 0.6%Ag and very low Si and Fe, with a UTS of 500 MPa.

Finally, for high wear specifications, 390 alloy containing 17%Si may be used [9], but this alloy has very low ductility and low impact strength.

Further information on the industrial uses of these alloys may be obtained from the papers of Garat [3, 9].

Fatigue Properties of Thixoformed Al Alloys

A detailed investigation on fatigue properties has been carried out by Badiali [19] to determine the fatigue strength in A356-T6 manufactured by different casting routes (see Table 2) and the nature of the defect initiating the fatigue crack. It is clear from the table that the fatigue strength of thixoformings at 10^7 cycles is slightly greater than squeeze castings, and significantly better than gravity die castings. The origin of crack initiation was observed to be either pores or oxide film, both being present in thixoforgings, and the fatigue strength increases with smaller defect size as might

Table 1 Mechanical properties of thixoformable casting alloys

Reference	Source	Temper. (MPa)	Yield strength (MPa)	UTS (MPa)	% Elong	Comments: Billets from EMS route unless stated otherwise (HT: heat treatment)
A356 (0.3%Mg)						
Chiametta (1996) [1]	Stampal	T5	180	255	5–10	
Bergsma (1997) [2]	Oregon/ Ancona	T5	155	240	3–14	HT: 7 h, 177°C; low ductility associated with porosity
Garat (1998) [3]	Pechiney, MHD	T5	170	260	15	
Fink, Witulski (1998) [4]	EFU, MHD	T5	185–200	230–250	2–5	
Giordano (2000) [5]	Stampal	T5	170–195	260–285	8–12	HT: 6 h, 170°C
Jorstad (2000) [6]		T5	200	260	12	
Uggowitzer (2002) [7]	ETH/ Zurich	T5	168	266	9.9	HT; 4 h, 160°C
Typical Value		*T5*	190	260	10	
Chiametta (1996) [1]	Stampal	T6	240	320	12	
Wendinger (1996) [8]	SAG	T6	230	300	12	
Bergsma (1997) [2]	Oregon/ Ancona	T6	230	300	5–14	HT: 4 h, 529°C; 5 h, 154°C
Garat (2000) [9]	Pechiney	T6	230	300	14	HT: 10 h, 540°C; 6 h, 160/170°C
Niu (1998) [10]		T6		305	11	
Jorstad (2000) [6]		T6	240	315	12	
Fink, Witulski (1998) [4]	EFU, MHD	T6	250–270	290–320	5–12	
Juang (2000) [12]		T6	276	341	11.5	HT: (12 h, 540°C; 12 h, 155°C) specimens machined directly from EMS billet
		T6	280	343	10.8	HT: (12 h, 540°C; 19 h, 155°C)
Boero, Chiametta (2000)[13]	Stampal	T6	220–240	290–310	10–16	HT: 6 h,160°C
Chiametta (2002) [14]	Stampal	T6	229	299	12	

(continued)

Table 1 continued

Reference	Source	Temper. (MPa)	Yield strength (MPa)	UTS (MPa)	% Elong	Comments: Billets from EMS route unless stated otherwise (HT: heat treatment)
Uggowitzer (2002) [7]	ETH/Zurich	T6	260	320	10.6	HT: 12 h, 540° C; 4 h, 160° C
		T6×3	229	317	16.7	Silicon spheroidisation: 3 min at 540° C; 4 h, 160° C
Typical Value		*T6*	230 ± 10	320 ± 10	12	
Uggowitzer (2002) [7]	ETH/Zurich	T4×3	115	232	17.4	Silicon spheroidising treatment only
Chiametta (1996) [1]	Stampal	T6	260	330	9	
Windinger (1996) [8]	SAG	T6	280	330	8	
Bergsma (1997)[2]	Anconna/ Oregon	T6	260	330	4–12	HT:8 h, 538°C; 12 h, 154°C: SSTT – (low ductility is associated with porosity)
		T6	250	305	7–13	HT:8 h, 538°C; 12 h, 154°C: EMS – (low ductility is associated with porosity)
Garat (2000) [9]	Pechiney	T6	290	345	10	
Jorstad (2000) [6]	Formcast	T6	250	315	10	
Bergsma (2001) [15, 16])	Anconna/ Oregon	T6	300	341	11	HT: 3 h, 538°C; 7 h, 177°C?: EMS
Boero (2000) [13]	Stampal	T6	290–310	340–360	5–7	HT: 6 h,170°C
Typical Value		*T6*	290	340	10	
Al–7Si–0.8Mg (exptl.)						
Uggowitzer (2002) [7]		T5	195		7.8	
		T6	319		7.4	12 h, 540°C + 4 h, 160°C

(continued)

Table 1 continued

Reference	Source	Temper. (MPa)	Yield stress (MPa)	UTS (MPa)	% Elong	Comments: Billets from EMS route unless stated otherwise (HT: heat treatment)
C355 Al–6Si–1Cu–Mg						
Garat (1998) [3]	Pecheney	T5	225	320	7	
319 Al–6Si–3Cu–0.3Mg						
Garat (1998) [3]						
(2000) [9]	Pechiney	T5	225			
Bergsma (2001) [15, 16]		T6	320	405	5	
	Anconna/ Oregon	T5	224	330	5	
		T6	350	400	5–6	
A390 (17%Si)						
Garat (2000) [9]	Pechiney	T5		250	<0.2	
		T6	350	350	<0.2	
Kapranos (2000) [17]	Pechiney	F	215	271	1.04	15%Si
		T6		429	0.16	15%Si
201 Alloy						
Liu (2004) [18]	Sheffield	T6	450	500	6–11	CS route, HT: standard
A357 (0.5%Mg)						
Chiametta (1996) [1]	Stampal	T5	200	285	5–10	
Garat (1998) [3]	Pechiney	T5	210	275	10	HT: 6 h, 170°C
Jorstad (2000) [6]	Formcast	T5	210	290	10	HT: no details
Boero (2000) [13]	Stampal	T5	195–210	265–285	5–8	HT: 6 h, 170°C
		T5	190–200	270–285	4–9	HT: 2 h, 180°C
Bergsma (2001) [15, 16]	Anconna/ Oregon	T5	220	290	5.5	direct chill casting+ thermal trans; optimised HT: 8 h, 182°C
Typical Value		*T5*	200	280	8	

be expected. However, it was considered that oxide films have a less deleterious effect on fatigue life than shrinkage pores of similar size.

Bergsma et al. [15] have carried out some work on 357-T6 and 319-T6 alloys on fatigue behaviour up to 10^5 cycles and have shown that thixoformed 357 is superior to conventionally cast A356 alloy and that a modified 357 alloy (low in Fe, Mn and Ni) is significantly better than conventionally as cast 319. The fatigue results reported by Das Gupta [20] are also given in Table 2 on several thixoformed casting

Table 2 Fatigue properties of thixoformed Al alloys

Alloy	Reference	Yield strength (MPa)	Fatigue strength (MPa at 10^7 cycles)	Defect type (μm^2)	Defect area	Process
A356-T6	[18]	267	135	Oxide/ porosity	20,000	Semisolid
		267	130	Oxide	40,000	Squeeze casting
		234	111	Porosity	80,000	Gravity casting
356-T6	[19]	223	117			
357-T6		290	120			
319-T6		345	165			
201	[20]	460	120			

alloys in the T6 condition. Again excellent fatigue strength of 319 (low in Fe, Mn and Ni, and therefore free of inter-metallic impurities) at 10^7 cycles is revealed, although a somewhat lower strengths are reported for 356 and 357 compared to Badliani.

Toughness and Wear Resistance

There is little information on impact strength or fracture toughness of the above casting alloys as thixoformed components. The available data (Das Gupta [20]) indicates that the toughness of A357/357 are similar to those obtained in squeeze castings in the T6 condition, and that the values for A356 and 319 are clearly superior to 357. The wear properties of SSM alloys are much improved over gravity permanent mold castings, and 390 thixoformed alloy is significantly better than all other SSM alloys and conventional die castings.

Conventionally Wrought Alloys

The most striking feature of Table 3 is the disparity in the results of mechanical properties within each alloy. The difficulty in making sense of these is that there are so many variables that can affect the properties, apart from errors in measurement themselves which are not controlled or even recorded in most cases. Like the casting alloys, possible sources of variation are alloy composition, the method of slurry preparation, semisolid heating practice; then the different forming processes (injection under pressure into a closed die – thixocasting, or simply squeezing between dies – thixoforging); and finally the heat treatment given to the specimens subse-

Table 3 Mechanical properties of usually wrought aluminum alloys, thixoforged and in the T6 condition

Source/Ref	Manu.Process, Grain Size	Yield Strength (MPa)	UTS (MPa)	% Elong	Heat treatment	Comments
6082 Alloy						
DIN Spec	Wrought	260	310	6		
Hirt (1994) [22]	GR	330	370	6-Jan	Short, 6 h	
Fink (1998) [4]	MHD, 100mm	256	284	12.6	T62	Cracks and pores observed
Kopp (2000) [23]	MHD	260	300	5	4 h 520,7 h 160	Thixocast
	MHD	100	155	20–25	4 h 520,7 h 160	Thixoforged
Engler (2000) [24]	MHD, 100mm	310	360	7.5	2 h 530, 6 h 160	With Si+Mg addn and no tears
Tausig (2000) [25]	GR, 140mm	201	249	16	Long soln treat	Thixoforged boss
		151	151	1	Long soln treat	Thixocast flange
Kaufmann (2000) [26]	NRC, 120mm	277	319	6.7	0.5 h 540, 10 h 170	
	NRC, 120mm	320	345	3.6	1.25 h 540,10 h 170	10 mm plate
	NRC, 120mm	355	383	2.4		Thin 6 mm plate
	NRC, 120mm	346	377	3.9		10 mm plate
Wabusseg (2002) [27]	NRC, 120mm	340	370	3.5		
	NRC, 120mm	320	350	3.6		0.2% B addn
Liu (2003) [28]	Extn, RAP	189	303	24	6 h 550, 2 h 180	0.1% Si
	MHD	231	302	12.3	6 h 550, 2 h 180	Low Fe, Mn, Cu
6062 Alloy						
Fink (1998) [4]	MHD	340	370	8		
7010 Alloy						
Lui (2003) [28]	CS	475	524	3.3	17 h 480, 4 h 120	

References	YS (MPa)	UTS (MPa)	%Elong.	Prodn.Route	Comments
7075 Alloy (Al–6Zn–2.5Mg–2Cu) in T6 condition					
Metals Handbook (1985) [29]	500	570	11		HT:Soln.465°C; 24 h,120°C

(continued)

Table 3 continued

References	YS (MPa)	UTS (MPa)	%Elong.	Prodn.Route	Comments
Kenney et al. (1998) [30]	421	496	7		
Hirt, EFU (1992) [31]	361	405	6.6		SIMA, 60–70 um particles
Rachmat. (2000) [32]	512	549	11	GR(60um)	moldless casting
Tausig (2000) [24]	431	465	4	low superheat	grain size ∼150 um
Chayong (2002) [33]	475	520	7		
Liu (2003) [28]	420	522	13.4	RAP	HT:17 h, 480°C; 4 h, 120°C
	397	486	8.8	CS	HT:17 h, 480°C; 4 h, 120°C
	441	538	6.8	CS	HT:17 h, 480°C; 4 h, 120°C*
	499	556	4	CS	HT:17 h, 480°C; 34 h, 120°C* * modified alloy containing 4Mg and low Fe, Mn, Si
2014 Alloy (Al–4Cu–1Si–0.5Mn) T6 condition					
Metals Handbook (1985) [29]	414	483	13	Wrought	
Liu (2003) [28]	219	283	3.6	RAP	1 h, 495°C; 8 h, 161°C
	314	356	2.9	RAP	1 h, 495°C; 18 h, 161°C
	399	474	3.6	RAP	17 h, 495°C; 24 h, 160°C
	393	454	6.3	RAP	17 h, 500°C; 24 h, 160°C
	270	408	19.6	CS mod.*	17 h, 500°C; 24 h, 160°C * 0.14%Si; low Fe, Mn, Zn
2024 Alloy (Al–4.4Cu–1.4Mg–0.6Mn)					

(continued)

Table 3 continued

References	YS (MPa)	UTS (MPa)	%Elong.	Prodn.Route	Comments
Metals Handbook (1995) [29]	395	476	10	Wrought	
Tietmann (1992) [31]	277	366	9.2	RAP	
Rachmat (2000) [32]	434	509	2.4	GR	Pressed, not thixoformed
Cho (2000) [34]	236	387	21	RAP	
Metal Matrix Composites					
Fink/Witulski (1998) [4]	410	646	4		T4; 2124+25%SiC
	285	2.5	2.5		T6; A356+16%SiC

GR: Grain Refined
CS: Cooling Slope

quent to the forming operation to raise the tensile strength, involving solution and ageing treatments.

It is believed (see Campbell [35]) that the yield stress (0.2% proof stress) of heat treated aluminium alloys is primarily due to the fine precipitates formed during the ageing treatment, and is essentially unaffected by most defects, such as porosity or coarse intermetallic phases. This would suggest that the variation in yield strength evident in the Table for a given alloy arises either from variation in alloy composition or to different heat treatments. The effect of different heat treatments is clearly demonstrated in the work of Liu et al. [28] using the same extruded bar of 2014 alloy. It is seen first that increasing the ageing time from 8 to 18 h increases the yield stress from 219 to 314 MPa, and furthermore increasing the solutionising time 1–17 h followed by ageing for 24 h at 160°C (a conventional ageing treatment) raises the yield stress to almost 400 MPa, close to the accepted value for the conventionally wrought alloy in the T6 condition. The longer solutionising time required in the thixoformed specimens no doubt reflects a coarser microstructure and underlines the need to develop new guidance on heat treatment to produce optimum properties for these particular alloys. A similar story may be seen in their work on the 7075 alloy: Ageing for 34 h at 120°C results in a substantial improvement in yield stress over a 4 h treatment, again resulting in a maximum yield stress of 500 MPa, the accepted T6 value recorded in the Metals Handbook [29].

The variation in yield strength evident in the table for the 6082 alloy, ranging from 260 to as high as 350 MPa can most probably be ascribed again to different heat treatment. Uggowitzer et al. [27] achieved the highest value using a 5 h solutionising treatment at 540°C followed by the long ageing of 10 h at 170°C, but also showed that the range of strengths observed in this same alloy can be obtained by varying the homogenisation time at 540°C (see table). However, it is also possible that the strength is affected by the local alloy composition being altered by segregation effects: this is the explanation given for the low value recorded by Kopp [23] in the thixoforged section of a complex shaped component.

However, it is also clear that these high strengths in all these alloys are achieved at the expense of ductility. Most researchers ascribe failure in thixoformed alloys to the presence of hot tears or porosity formed during solidification, which can grow under the right stress condition. For the same defect size and population, these tears or pores will grow rapidly in high yield strength alloys with little deformation; whereas for low strength alloys, significant deformation and work-hardening is required to raise the local stresses to a level where crack propagation will occur rapidly. It is to be expected therefore that increasing the strength by ageing also leads to reduced ductility.

Exceptions to this rule are seen in the work of Rachmat et al. [32] in 7075, where both high strength and ductility (11%) were achieved "in the fully heat treated condition" (undefined) and to a lesser extent by Tausig [25] (16% elongation) in 6082. The 7075 specimens were grain refined ($\sim$ 60 μ m grains) and essentially forged in the semisolid state between two heated platens, as was the 6082 alloy. In this situation, the slurry is essentially under hydrostatic pressure that does not allow the formation of pores or hot tears. It is of no doubt that the fine grain size permits easy flow to accommodate solidification shrinkage. The resulting essentially defect-free microstructure is believed to be responsible for the enhanced ductility under these processing conditions.

In conclusion, it seems that high strengths can be obtained in these precipitation-hardening alloys after thixoforming and which, given the appropriate heat treatment, are close to the values traditionally achieved in the wrought alloy. However, defects generated during solidification of the semisolid alloy, such as shrinkage pores or hot tears, limit the ductility. It might be possible to overcome this to some extent by attention to die design in order to allow mass feeding during the critical cooling stages, and perhaps by tailoring the alloy composition. For safety-critical parts of complex shape and high value, such as required in the air industry, hot isostatic pressing (hipping) [35] might be justified to reduce or eliminate such defects and improve their impact strength.

Metal Matrix Composites

An early interest was shown in producing MMC components by the semisolid route, in particular, using an aluminium alloy matrix in which Al_2O_3, SiC and TiC particulates were dispersed either by stirring into the semisolid melt (compocasting), by injecting into an alloy spray (Osprey process) or by the more conventional method of mixing powders. Despite the feasibility of such processing and the fundamental work on the rheological behaviour of these slurries (see section "Influence of Shear Rate" in Part II, Modeling and Rheology) and the impressive strength properties obtained in the T4 condition [4], little commercial interest has been maintained so far in these products because of their low ductility and the expense of multi-step processing. Hypereutectic Al–Si alloys, however, may be regarded as natural MMCs containing hard refractory silicon particles and considerable interest is still evident in this alloy series, despite again their poor ductility and the efforts to reduce particulate size to overcome this limitation.

References

1. Chiametta, in *Proceedings of the 4th S2P International Conference*, Sheffield, 1996, pp. 204–207
2. Bergsma, Tolle, Kassner, Li, Evangelista, Mater. Sci. Eng. **A237**, 24–34 (1997)
3. Garat, in *Proceedings of the 5th S2P International Conference*, Golden CO, 1998, pp. 199–213
4. Fink, Witulski, in *Proceedings of the 5th S2P International Conference*, Golden CO, 1998, pp. 557–564
5. Giordano, Boero, Chiametta, in *Proceedings of the 6th S2P International Conference*, Turin, 2000, pp. 29–34
6. Jorstad, in *Proceedings of the 6th S2P International Conference*, Turin, 2000, pp. 227–233
7. Orgis, Luchinger, Uggowitzer, in *Proceedings of the 7th S2P International Conference*, Tsukuba, 2002, pp. 713–718
8. Wendiger, in *Proceedings of the 4th S2P International Conference*, Sheffield, 1996, pp. 239–241
9. Garat, Maenner, Sztur, in *Proceedings of the 6th S2P International Conference*, Turin, 2000, pp. 186–194
10. Niu, Hu, Hao, Yec, Pinwell, in *Proceedings of the 5th S2P International Conference*, Golden CO, 1998, pp. 141–148
11. Sukumaran, Pillai, Ravikumar, Praveen, Kelukutty, Soman, in *Proceedings of the 5th S2P International Conference*, Golden CO, 1998, pp. 379–385
12. Juang, Wu, Ma, Peng, in *Proceedings of the 6th S2P International Conference*, Turin, 2000, pp. 705–710
13. Boero, Chiametta, Giordano, *Proceedings of the 6th S2P International Conference*, Turin, 2002, pp. 581–586
14. Rosso, Romano, Giodano, Chiametta, *Proceedings of the 7th S2P International Conference*, Tsukuba, 2002, pp. 151–156
15. Bergsma, Li, Kassner, Mater. Sci. Eng. **A297**, 69–77 (2001)
16. Bergsma, Kassner, Evangelista, Cerri, *Proceedings of the 6th S2P International Conference*, Turin, 2000, pp. 319–324
17. Kapranos, Kirkwood, Atkinson, Rheinlander, Bentzen, Toff, Debel, Laslaz, Maenner, Blais, Rodriguez-Ibabe, Lasa, Giordano, Chiametta, *Proceedings of the 6th S2P International Conference*, Turin, 2000, pp. 741–746
18. Liu, Atkinson, Kapranos, Jones, J. Mater. Sci. **39**, 99–105 (2004)
19. Badiali, Davidson, Griffiths, Zanada, *Proceedings of the 6th S2P International Conference*, Turin, 2000, pp. 349–354
20. Das Gupta, *Science and Technology of Semisolid Metal Processing*, chap. 3 (Worcester Polytechnic Institute, Worcester, MA)
21. Kapranos, Apt Aluminium, Process. Product Technol. **Feb**, 39–44 (2008)

22. Hirt, Cremer, Winkelmann, *Proceedings of the 3rd S2P International Conference*, Tokyo, 1994, pp. 107–116
23. Kopp, Winning, Kallweit, Knissel, Gabathuler, Quast, *Proceedings of the 6th S2P International Conference*, Turin, 2000, pp. 687–691
24. Engler, Hartmann, Niedick, *Proceedings of the 6th S2P International Conference*, Turin, 2000, pp. 483–488
25. Tausig, *Proceedings of the 6th S2P International Conference*, Turin, 2000, pp. 489–494
26. Kaufmann, Wabusseg, Uggowitzer, *Proceedings of the 6th S2P International Conference*, Turin, 2000, pp. 457–462
27. Wabusseg, Gullo, Uggowitzer, J. Mater. Sci. **37**, 1173–1178 (2002)
28. D. Liu, Atkinson, Kapranos, Jirattiticharoean, Jones, Mater. Sci. Eng. **A361**, 213–224 (2003)
29. Metals Handbook (1985)
30. Kenney et al., *Metals Handbook*, 9th edn., vol. 15, pp. 327–338
31. Tietmann, Bremer, Hirt, Kopp, *Proceedings of the 2nd S2P International Conference,* pp. 170–179
32. Rachmat, Takano, Ikeya, Mater. Sci. Forum **329–330**, 487 (2000)
33. Chayong, Atkinson, Kapranos, Mater. Sci. Eng. **A390**, 3–12 (2005)
34. Cho, Kang, J. Mater. Proc. Technol. **105**, 269 (2000)
35. J. Campbell, *Castings* (Butterworth Heinemann, Germany, 1991)

Index

GPSR Compliance

The European Union's (EU) General Product Safety Regulation (GPSR) is a set of rules that requires consumer products to be safe and our obligations to ensure this.

If you have any concerns about our products, you can contact us on ProductSafety@springernature.com

In case Publisher is established outside the EU, the EU authorized representative is:

Springer Nature Customer Service Center GmbH
Europaplatz 3
69115 Heidelberg, Germany

Batch number: 10584666

Printed by Printforce, the Netherlands